Heat Transfer

A Laboratory Manual

(Under graduate Chemical Engineering)

N S Srinivas

Blank Page

Dr N S Srinivas (Retired)   email ID: nssrinivas602@gmail.com

 Assistant Professor

Department of Chemical Engineering

Indian Institute of Technology, Madras

Chennai, India

Assistant General Manager

Boiler Division

Walchandnagar Industries Ltd

Pune , India

Principal

Sri Venkateswara Engineering College for Women

Tirupati

Present Address

2/4 Amolenclave, Kotkar line

Bhaupatil Rd, Pune  411020

Maharashtra, INDIA

The author has taken his best effort in the accuracy and method of calculations for each of the experiment. The drawing are made in Powerpoint and pasted in the word document. The source code for all the experiments in C++ language which is tested by the author is presented at the end of the experimens. The author and publisher offer no warranties or representations, nor do they accept any liabilities with respect to the use of this information. Please report any errata to the author.

BlankPage

## Preface

This laboratory manual is the second edition of the previously published by the "Chemical Engineering Education Development Centre" of Indian Institute of Technology, Madras, Chennai 600036. in the year 1980, February.

Wide rages of experiments to cover the principle aspects of heat transfer operations of interest to engineers in general and to Chemical Engineers in particular are included in this manual. Typical experimental units are described here but individual laboratories may modify these or have alternative units.

Usually students may have to carry a latter experiment, even before the relevant theory portion is covered in the class. Therefore each experiment is presented such that the reference to an earlier experiment is not needed to carry out a latter experiment.

This laboratory manual is not intended as a substitute for record writing but to ensure that experiments are carried out with understanding, experimental data are recorded correctly and results are calculated and presented in a logical manner.

It is advisable that students should develop the attitude of consulting .text books and hand books for collection and checking the data.

For the calculations of the experimental data, the author has given tested source code for all the experiments in C++ language.

N S Srinivas

Pune
April 2014

Blank page

CONTENTS

|  |  | page |
|---|---|---|
| | Preface | v |
| | Introduction | ix |
| 1. | Determination of thermal conductivity | 1 |
| 2. | Insulation thickness | 9 |
| 3. | Electrical analogue : steady state conduction heat transfer | 19 |
| 4. | Unsteady state heat transfer | 25 |
| 5. | Effective thermal conductivity of a packed bed | 31 |
| 6. | Heat transfer by free convection | 35 |
| 7. | Double pipe heat exchanger | 41 |
| 8. | Finned tube heat exchanger | 49 |
| 9. | Shell and tube heat exchanger | 57 |
| 10. | Heat transfer in agitated vessels | 63 |
| 11. | Heat transfer to boiling liquids | 69 |
| 12. | Heat transfer in gas fluidized beds | 75 |
| 13. | Long tube vertical evaporator | 81 |
| 14. | Radiation constant | 88 |
| 15. | Source code in C++ | 93 |
| | References | 145 |
| | Appendix- Factors for conversion to SI units | 147 |

Blank page

## Introduction

A laboratory course helps the student to verify and satisfy himself by experimentation the various fundamentals he has come across the lecture class. He can verify the results obtained by experimentation by comparing with those obtained by calculations using empirical relation; thereby he gets an idea of the limitations of the empirical expressions.

Many of the experiments cannot be done by a single individual. The students work in groups of two or three. Team work develops interpersonal skills, such as communication, planning, working in as group and decision taking.

The main advantage of a laboratory course is that the ideas gained by performing the experiments in the laboratory are more vivid and lasting. Hazy notions are dispelled when a student sees for on his own the facts unfold themselves before his eyes. He comes to recognize what is important and what can be neglected. Finally he learns through his mistakes, and if he is persistent enough, can narrow down the disparities between the experimental and theoretical results.

## WHAT IS EXPEXTED OF A STUDENT IN THE LAORATORY

Before starting the experiment straight away, the experimenter must make a through study of the experimental set up. He should know how the unit works, what are the functions of various valves, types of valves used and the reason for the same, and which valve controls which stream. He must also have a prior knowledge of the limitations under which the unit can work (For instance, if steam is used as a heating medium what is the maximum pressure that can be used safely?). Similarly the limitations of a voltmeter or ammeter should be noted. While preparing experiments to estimate pressure drop due to flow of fluids in conduits, manometers are used. There may be manometer containing a liquid of low density like carbon-tetrachloride to determine pressure drop at low flow rates or a mercury manometer for high flow rates. The experimenter must find out at the outset up to what range he can use safely the carbon-tetrachloride manometer and when he should switch over to the mercury manometer.

After making through study of the experiment the experimenter must write down in what sequence the experiment is to be performed and what data to be collected. This helps the experimenter to see if he has missed any data however trivial it may look. Many a time, students determining heat losses from a lagged pipe may forget to take the ambient temperature. This may particularly be bottle neck for calculations later, because the ambient temperature changes from time to time.

After the day's work the experimenter should close all the valves, switch off motors etc., and keep the unit ready to be operated by the other students in the next class. Stop watches, thermometers, voltmeters, ammeters etc., should be returned to the lab-in-charge.

The observations made, data collected and the results must be presented in the form of a report. The report must include the following;

(1) Introduction: This must include object of the experiment, why it is being done and what would be the applications of the results and conclusions.

(2) Theory : A brief statement of the theory which forms the background of the experiment must be written giving all pertinent equations and expressions that will be used. These need not be derived but the source of information must be indicated.

(3) Experimental procedure: A line diagramof the entire set-up showing the main unit, the valves, pumps, pressure gauges, electrical connections and other measuring instruments must be drawn. A description of the main apparatus and a brief outline of procedure indicating the sequence of operations must be given.

(4) Final results: They are to be presented in tabular form and/ or graph.

(5) Discussion of the results and conclusions: this is the vital part of the report. Any deviations of the results are to be noted and analyzed. The cause for any abnormal data must also be discussed. Any auxiliary problem which arises from the experiment and which can be investigated with minor modifications of the main unit also find place.

(6) Appendix: This most include the data sheet containing all the data collected while performing the experiment and sample calculation.

The report must be so written that it is easy to understand.

## LABORATORY EXCERCISES IN HEAT TRANSFER

The laws of heat transfer are very important in the design and operation of many industrial types of equipment like, pre-heaters, exchangers, condensers, etc. The primary object is to achieve maximum heat transfer rate per unit surface.

The most common equipment in industry is a heat exchanger, where process streams are either cooled or heated. The design of heat exchanger requires the calculation of flow coefficients of heat transfer for the process streams, and conductive heat transfer through the metal wall that separates the two fluids. Apart from this, the heat exchangers loose heat to the surroundings; in order to minimize this, the surface of the exchanger xi

must be well insulated. Even in spite of insulation, a steam pipe loses heat to surroundings from the insulated surface by natural (or free) convection and radiaton.

The design of a boiler or condenser requires the knowledge of both boiling and condensation heat transfer coefficients.

This manual is written with the aim that a laboratory course helps the student to verify the experimental results using the existing correlations.

The experiments described in this manual use water, air and steam. Every set-up has been treated as a separate unit with separate tanks, and steam in-let. However, a central tank with separate pumps may be used for individual set-up if required.

Blank Page

# 1. Determination of thermal conductivity of metals

**Nomenclature**

A    Area of cross section, $m^2$

$C_p$    Specific heat, J/kgK

d    Diameter of the rod, m

k    Thermal conductivity, W/mK

L    Length of rod, m

m    Mass flow rate of fluid, kg/s

Q    Rate of heat transfer, W

q    Heat flux, $W/m^2$

r    Radial direction, m

T    Temperature, $^{\circ}C$

$T_{in}$    Temperature of inlet water, $^{\circ}C$

$T_{out}$    Temperature of outlet water, $^{\circ}C$

w    Condensate rate, kg/s

$\Delta X$    Distance between the planes, m

$\lambda$    Latent heat of vaporization, J/kg


Subscripts

1    plane 1

2.    plane 2

x, y, z In directions of x, y and z coordinates


## 1.1 Introduction

Heat is a form of energy which flows from hotter to a colder part of a substance or from a body at a higher temperature to one at a lower temperature. The rate of heat flow depends mainly upon the physical properties and temperatures of the bodies and the ambient conditions. When two ends of an iron rod are held at two different temperatures, heat flows from the warmer end to the cooler end, though the substance itself does not move. This process of heat transfer by conduction arises from transfer of free electrons and the transfer of vibration energy from one molecule to another.

The thermal conductivity is a physical property which indicates the effectiveness of a substance in transferring heat by conduction. If the two faces of a body of thickness $\Delta x$ (Fig 1.1) are kept at $T_1$ and $T_2$, the heat flow $Q_x$ through the area A per unit time is given by

$$Q_x = k\,A\,(T_1 - T_2)\,/\,\Delta X \qquad\qquad 1.1$$

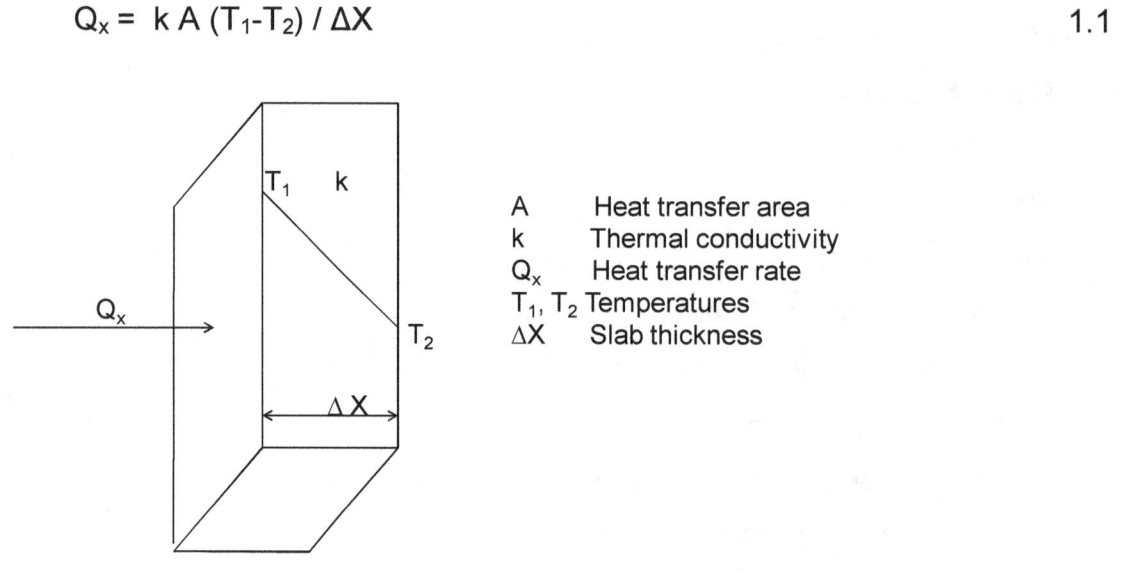

A     Heat transfer area
k     Thermal conductivity
$Q_x$     Heat transfer rate
$T_1$, $T_2$ Temperatures
$\Delta X$     Slab thickness

Fig 1.1 Conduction heat through a wall

In Eq.(1.1), $Q_x$ is the heat transferred per unit time in x-direction and k is the thermal conductivity. In general, thermal conductivity is a function of temperature and the mean value over the temperature range $T_1$ and $T_2$ used in equation (1.1).

It will be more convenient to write equation (1.1) in differential form to use in analytical expression. For this, one uses the limiting form of this equation, as the slab thickness $\Delta x$ approaches zero.

The local heat flow per unit area (heat flux) in the positive x-direction is designated by $q_x$. In this notion, equation (1.1) becomes.

$$q_x = Q_x / A = -k \, dT/dx \qquad\qquad 1.2$$

Equation (1.2) is the one dimensional form of Fourier's law of heat conduction, valid when $T = T(x)$. For an isotropic medium in which the temperature varies in the three directions, we can write similar equations for each of the coordinate directions, viz.

$$q_x = -k \, dT/dx$$
$$q_y = -k \, dT/dy \qquad\qquad 1.3$$
$$q_z = -k \, dT/dz$$

These three relations given in equation (1.3) are the components of a single vector equation given by

$$\vec{q} = -k \, \nabla T \qquad\qquad 1.4$$

Which is three dimensional form of Fourier's law and states that the heat flux.

Vector $\vec{q}$ is proportional to the temperature gradient $\nabla T$ and is oppositely directed.

## 1.2 Objective of the experiment

To determine the thermal conductivity of copper

## 1.3 Theory

Consider a copper rod of diameter d and length L whose ends are maintained at different temperatures. The rod is provided with thermocouples at equal distances along the axis of the copper rod to find out the temperature profile along the axis. One end is connected to a steam chest and the other end to a well-stirred water tank which removes heat. The rod is well insulated to minimize heat loss to the surroundings from the rod surface. In the present case, heat flow is assumed to take place only along the axis of the rod and the radial temperature gradient viz., (dT/dr) is neglected.

From the thermocouple readings the temperature T of the rod is obtained as a function of distance along the length of the rod. Heat transfer from the rod to water is given by

$$Q_x = q \times A = A \left( -k \, dT/dx \big|_{x=L} \right) \qquad\qquad 1.5$$

Where $A = \pi \, d^2/4$

and $dT/dx \big|_{x=L}$ is temperature gradient at the end of the rod

### 1.3.1  Heat balance around the water tank

Here an assumption is made regarding the outlet temperature of the water. The water leaving the tank is assumed to have the same bulk temperature as the water in the tank, since the liquid is well stirred due to which temperature gradients cease to exist, that is uniform temperature through out.

At steady state, the heat taken up by the water must be equal to that given by the rod at $x = L$. Thus,

Heat taken up by water $(Q_w) = m \, C_p \, (T_{out} - T_{in})$ $\qquad\qquad 1.6$

This heat balance equation at steady state becomes

$$Q_w = -A \, k \, dT/dx \big|_{x=L} = m \, C_p \, (T_{out} - T_{in}) \qquad\qquad 1.7$$

Therefore

$$k = m C_p (T_{out} - T_{in}) / A ( - dT/dx |_{x=L}) \qquad 1.8$$

Heat loss from the rod through insulation, which cannot be eliminated absolutely can be estimated from steam condensate rate as,

$$Q_{loss} = Q_x |_{x=0} - Q_x|_{x=L} \qquad 1.9$$

Where $Q_x|_{x=0} = w\ I$

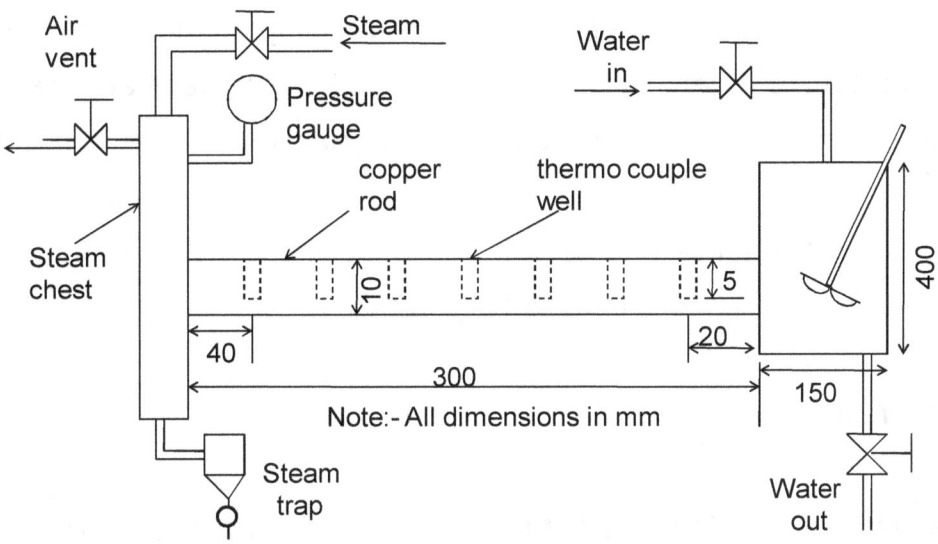

Fig 1.2 Experimental set-up

## 1.4    Experimental set-up

A schematic presentation of experimental set-up is given in Fig 1.2. A copper rod of 10 mm in diameter and 300 mm length is taken as the experimental piece. The copper rod is provided with 7 iron-constantan thermocouples (30 S.W.G.). The thermocouples are soldered with soft lead to ensure good contact between the bead of the thermocouple and the copper road. The cold junction is maintained at $0^\circ C$ and potential differential is measured with a millivoltmeter having 10 micro volts resolution. These thermocouples are already calibrated before they are soldered. The entire length of the rod is insulated to minimize the heat losses to the surroundings. Water flow rate is measured with a Rotameter. The steam chest is provided with a pressure gauge, air vent and a steam trap.

## 1.5    Experimental Procedure

Steam pressure about150 $kN/m^2$ is maintained in the steam chest. The cooling water flow rate to the tank is set to a particular value. After the system reaches steady state conditions as indicated by water out let temperature, the readings of the thermocouples attached to the copper rod, water temperature in the tank, inlet water temperature, water flow rate and steam condensate rate are measured. In order to find the temperature gradient $(dT/dx)|_{x=L}$ ( ie., at the end of the rod), a plot of T, (the temperature of the rod by the thermocouples) against distance x is drawn. The experiment is repeated for three flow rates of water keeping the steam at constant pressure.

## 1.6    Presentation of data

Note the thermocouples readings from the hot end for distances of 5, 10, 15, 20, 25 and 30 cm. and also the steam pressure in $kN/m^2$

Tabulate the following

(1) Mass flow rate of water, kg/s

(2) Condensate rate, kg/s

(3) Temperature of out let water, $^\circ C$

(4) Temperature of inlet water, $^\circ C$

(5) Temperature gradient, $dT/dx|_{x=L}$, $^\circ C/m$

Tip: water or condensate is collected in a known volume container for a fixed time with the help of a stop watch. Rate of flow is the volume of the collected divided by the collection time of either water or condensate.

## 1.7    Results

Plots of $mC_p$ ($T_{out}$ -$T_{in}$) vs., A (-$dT/dx|_{x=L}$) are drawn for different cooling water flow rates and steam pressures. The value of k, the thermal conductivity of the material is determined from equation (1.8).The value of k, thus determined shall be compared with the value reported in page 23-48, Chemical Engineer's Hand book $5^{th}$ ed.

Note: - Instead of cooper rod, any other material such as brass, or stainless steel of same dimension may be used

The condensate temperature is related to the saturated steam pressure. One can choose about three to four steam pressures.

## 1.8    Questions

1) Define Fourier's law
2) Define isotropic and non-isotropic property
3) From the following figure state which material has high thermal conductivity? Explain why?

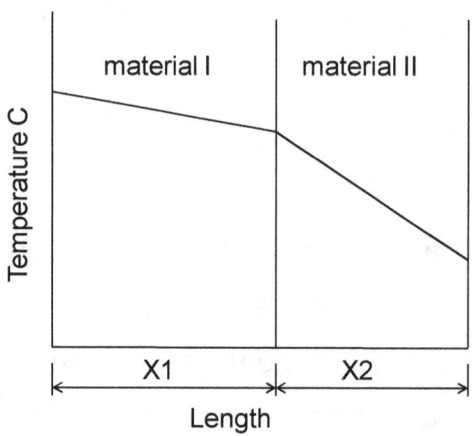

Figure for question 3

4) Explain the term steady state.

## 1.9    Data collection

| | | | 150 | 200 | 250 | 300 |
|---|---|---|---|---|---|---|
| A | Steam pressure, $kN/m^2$ | | 150 | 200 | 250 | 300 |
| B | Water quantity, kg | | | | | |
| C | Time taken for water , sec. | | | | | |
| D | Water flow rate , kg/s | B/C | | | | |
| E | Condensate quantity, kg | | | | | |
| F | Time taken for condensate, sec | | | | | |
| G | Condensate rate,  kg/s | F/G | | | | |
| H | Temperature of inlet water, $^{\circ}C$ | | | | | |
| I | Temperature of out let water, $^{\circ}C$ | | | | | |
| J | Temperature difference, $^{\circ}C$ | I - H | | | | |
| K | Distance between thermocouples, m | | | | | |
| L | Temperature gradient, $^{\circ}C$ /m | J/K | | | | |

# 2. Insulation Thickness

**Nomenclature**

$C_{pf}$   Specific heat of air, J/kgK

$g$        Acceleration due to gravity, m/s$^2$

$D_s$      Diameter of pipe with insulation, m

$h_a$      Natural convection heat transfer coefficient, W/m$^2$K

$h_f$      Inside film heat transfer coefficient, W/m$^2$ K

$h_o$      Outside film heat transfer coefficient, W/m$^2$K

$k_1$      Thermal conductivity of metal, W/mK

$k_2$      Thermal conductivity of insulation, W/mK

$k_f$      Thermal conductivity of air, W/mK

$L$        Length of the pipe, m

$Q$        Rate of heat transfer, W

$R_1$      Inner radius of metal pipe, m

$R_2$      Outer radius of pipe, m

$R_3$      Outer radius of pipe with insulation, m

$R_c$      Critical radius of insulation from pipe center, m

$T_i$      Steam temperature, °C

$T_s$      Surface temperature, °C

$T_\infty$ Ambient temperature, °C

$U_i$      Overall heat transfer coefficient based on inside area, W/m$^2$K

$W$        Condensate rate, kg/s

Dimensional groups

$N_{Gr}$ = Grashoff Number, $[ D_s^3 \rho_f^2 g \beta \Delta T/\mu_f^2]$

$N_{Nu}$ = Nusselt Number, $[h_a D_s / k_s ]$

$N_{Pr}$ = Prandtl Number, $[C_{pf} \mu_f / k_f]$

Greek letters

σ   Stefan-Boltzmann constant, $5.67 \times 10^{-8}$, $W/m^2K^4$

ΔT  Temperature difference ($T_s$ - $T_\infty$ ), $^oK$

λ   Latent heat of vaporization, J/kg

$\rho_f$   Density of air, $kg/m^3$

β   Coefficient of cubical expansion, 1/ $^oK$ (for gases)

$\mu_f$   Viscosity of air, $Ns/m^2$

ε   Emissivity

## 2.1 Introduction

A hot reactor, storage vessel or a steam pipe will lose heat to the atmosphere, if it is un-lagged or un-insulated. The heat loss may be due to radiation, conduction and convection. The loss of heat due to radiation is proportional to the fourth power of the absolute temperature of the body and the surrounding temperature. Further, it will be small for lower temperature differences but will increase rapidly as the temperature difference between the hot body and cold body increases. Since air has a low a lower thermal conductivity, the heat loss by conduction will be small. On the other hand, since convection currents form easily around the tube surface, the heat loss from an un lagged surface is considerable. As heat is to be conserved, hot surface should be suitably lagged or insulated. The insulation increases the resistance to heat flow and hence decreases the heat loss from the hot fluid within the pipe. More over if a pipe is very heavily insulated, the heat loss may again increase since the effect of increased heat transfer surface offsets the advantage of insulation, and thus there is an optimum or critical thickness of insulation which is a function of the thermal conductivity of the insulating material and outside convective heat transfer coefficient.

## 2.2 Objectives of the experiment

(a) To determine experimentally the critical radius of insulation over a copper pipe.

(b) To determine experimentally the overall heat transfer coefficient based on inside surface.

## 2.3 Theory

Consider a pipe of inner radius $R_1$ and outer radius $R_2$ which is insulated up to radius $R_3$. Fig 2.1 represents the cross-section of the insulated pipe. Under steady state heat transfer conditions, the rate of heat loss over a length L, of pipe is given by

$$Q = \frac{2 \pi L (T_i - T_\infty)}{1/h_i R_i + (1/k_1) \ln (R_2/R_1) + (1/k_2) \ln (R_3/R_2) + (1/h_o R_3)} \qquad 2.1$$

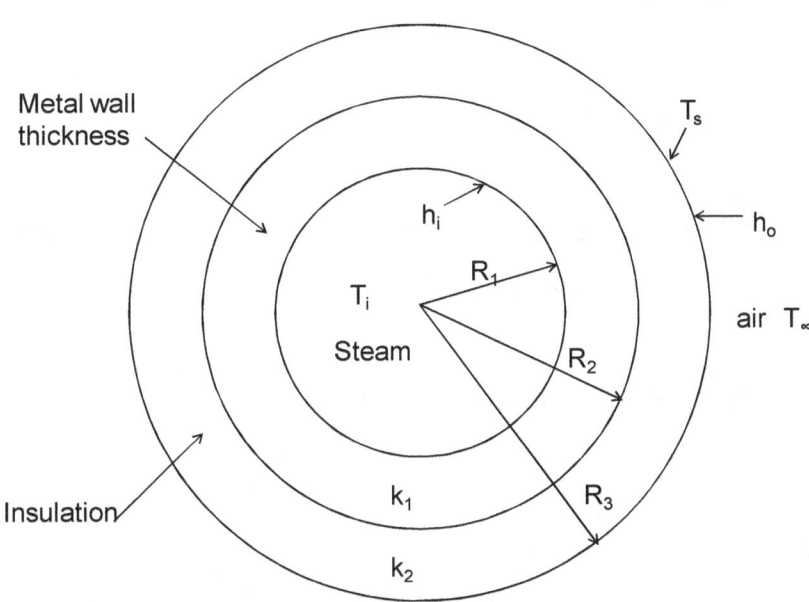

Fig. 2.1 Cross section of an insulated pipe

For a set of fixed values of $h_o$, $k_1$, $k_2$, there is a value of $R_3$ which will cause maximum heat loss. This is called critical radius $R_c$. This can be evaluated by differentiating equation Eq (2.1) with respect to $R_3$ and setting the result equal to zero. Thus

$$dQ/dR_3 = \frac{- 2 \pi L (T_i - T_\infty)[1/k_2R_2 - 1/h_o R_3^2]}{(1/h_i R_i + (1/k_1) \ln(R_2/R_1) + (1/k_2)\ln(R_3/R_2) + (1/h_o R_3))^2} = 0 \qquad 2.2$$

$$\text{or} \qquad R_3 = R_c = k_2/h_o \qquad\qquad\qquad 2.3$$

Equation (2.3) gives the critical radius of insulation (radius of insulation for which the heat losses become maximum) or critical insulation thickness is $(R_c - R_2)$. In equation (2.3) the out side heat transfer film coefficient is due to natural convection and radiation heat transfer.

Thus $h_o = h_a + h_r$ $\qquad\qquad\qquad\qquad\qquad 2.4$

The value of $h_a$ can be computed from the correlation.

$$h_a D_3/k_f = b [(D_3^3 \rho_f^2 g \beta \Delta T /\mu_f^2)(C_{pf} \mu_f/k_f)]^n \qquad\qquad 2.5$$

Where $k_f$, $\rho_f$, $\beta$, $\mu_f$ and $C_{pf}$ are properties of air and are evaluated at the arithmetic mean of the surface temperature of pipe / insulation and bulk air temperature.

The dimensionless terms appearing in equation (2.5) are

Nusselt number $\quad N_{Nu} \quad = h_a D_3/k_f$

Grashoff number $N_{Gr} \quad = D_3^3 \rho_f^2 g \beta\Delta T /\mu_f^2$

and

Prandtl number $\quad N_{Gr} \qquad = C_{pf} \mu_f/k_f$

Thus equation (2.5) can be written as

$$N_{Nu} = b \, [N_{Gr} N_{Pr}]^n \qquad\qquad 2.6$$

The constants b and n depend on the range of $N_{Gr} N_{Pr}$ and are summarized below

| $[N_{Gr} N_{Pr}]$ | b | n |
|---|---|---|
| $>10^9$ | 0.13 | 0.333 |
| $10^4 - 10^9$ | 0.53 | 0.25 |
| $<10^4$ | 1.09 | 0.125 |

The radiation heat transfer coefficient is given by

$$h_r = \sigma \, \varepsilon \, (T_s^4 - T_\infty^4) \, / \, (T_s - T_\infty) \qquad\qquad 2.7$$

The overall heat transfer coefficient based on inside area is given by

$$1/U_i = 1/h_i + (R_1/k_1) \ln (R_2/R_1) + (R_1/k_2) \ln (R_3/R_2) + (R_1/R_3 \, h_o) \qquad 2.8$$

In the present situation, the inside film coefficient and metal conductance are very large. So, the controlling thermal resistances to heat transfer are due to insulation and ambient air. Hence equation (2.8) can be simplified as

$$1/U_i = (R_1/k_2) \ln (R_3/R_2) + (R_1/R_3 \, h_o) \qquad\qquad 2.9$$

Normally the overall heat transfer coefficient will be based on the side where the film heat transfer coefficient is low. But due to the non availability of data, the coefficients were based on inside. However equation (2.9) can be modified with respect to the data available.

## 2.4 Experimental set-up

The experimental set-up is shown in Fig. 2.4. Five copper pipes (2.54 cm O.D. and 2 m long) are connected to a 4.0 cm. O.D. steam line having 10 cm. center-center axial distances. Each pipe is provided with a valve to adjust steam input, a pressure gauge to note steam pressure and a steam trap. Only four pipes are insulated with asbestos whose thickness is different over each pipe. Calibrated iron-constantan thermocouples (30 Gauge) are attached over the surfaces of insulation for each pipe (also on the bare pipe). Average of these thermocouple readings over each pipe is assumed to be the surface temperature of that particular pipe. The outer radius of each pipe's is more by 6 mm.

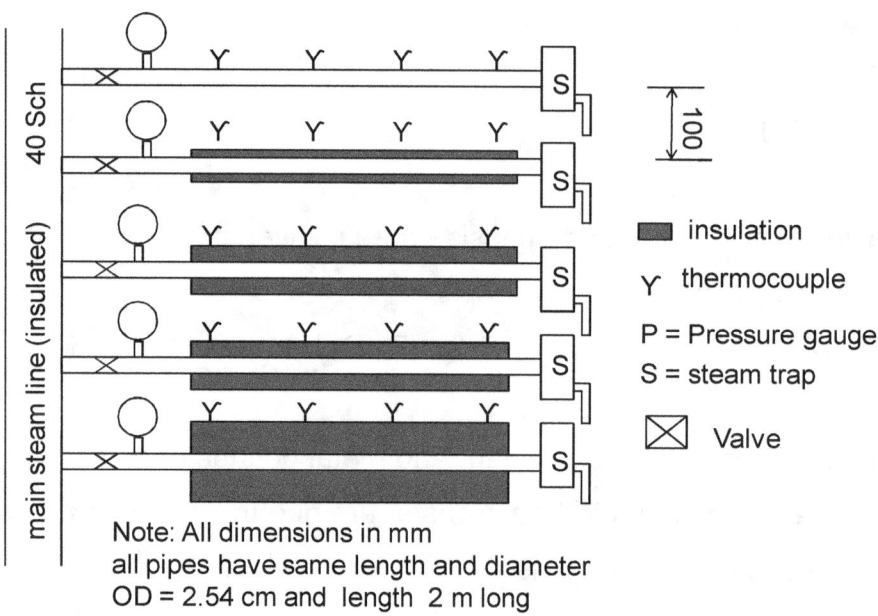

Note: All dimensions in mm
all pipes have same length and diameter
OD = 2.54 cm and length 2 m long

Fig. 2.2 Studies on insulation thickness

## 2.5 Experimental procedure

The effective lengths and outer diameters of all pipes are measured. Steam is admitted to the pipes and the pressure is recorded. After attaining steady state conditions as noted by thermocouples readings, the amount of condensate discharge is noted for 10 minutes from each pipe. The room temperature and the outer surface temperatures of the pipes are noted.

## 2.6 Presentation of data

Note the inner and outer radii of the pipes, length of the pipes.

Tabulate the following for each pipe

(1) Outer radius of insulation, m

(2) Steam pressure, $kN/m^2$

(3) Condensate collected in 10 minutes, kg

(4) Room temperature, $^{\circ}C$

(5) Average surface temperature, $^{\circ}C$

## 2.7 Results

The temperature and latent heat of saturated steam corresponding to the steam pressure is obtained from steam tables. From the amount of condensate discharged per unit time, the heat transfer rate Q can be calculated as

$$Q = w \lambda \qquad\qquad 2.10$$

The overall heat transfer coefficient based on the inside surface is given by

$$Q = U_i \, 2 \, \pi \, R_i \, L \, (T_i - T_\infty) \qquad\qquad 2.11$$

$T_i$ is considered as the steam condensing temperature at the steam pressure. The value of Q is calculated for each pipe from equation (2.10). The critical thickness corresponds to the pipe which gives the maximum value of Q.

Tabulate the following for each pipe

(1) Inside wall temperature, °C

(2) Latent heat of vaporization, J/kg

(3) Condensate rate, kg/s

(4) Heat loss to surroundings, W

(5) Overall heat transfer coefficient, $W/m^2 C$

In the plot of $R_3$ on x-axis, the value of $R_c$ corresponds to the maximum value of Q.

The experimental value of $k_2$ can be evaluated from equation (2.3) provided the value of $h_o$ is known. The value of $h_o$ can be calculated from equation (2.4) together with equations (2.5) and (2.7).

The values of $U_i$ calculated from equations (2.9) and (2.11) are to be tabulated as follows for comparison.

(1) $U_i$ from equation (2.11), $W/m^2 K$

(2) $h_a$ from equation (2.5), $W/m^2 K$

(3) $h_r$ from equation (2.7), $W/m^2 K$

(4) $h_o$ from equation (2.4), $W/m^2 K$

(5) $U_i$ from equation (2.9), $W/m^2 K$

The value of $k_2$, evaluated from equation (2.3) can be compared with the values of k for insulating materials reported in page 11-48 to 11-52, Chemical Engineer's Hand Book, 5ed.

Note: Instead of asbestos insulation any other insulation reported in Perry 5 th ed may be used.

## 2.8 Questions

(1) Explain the critical insulation.

(2) What is the significance of $N_{Gr}$ ?

(3) What is the signification of $N_{Pr}$ ?

(4) Show that $N_{Gr}$ and $N_{Pr}$ are dimensionless.

(5) Explain the purpose of a steam trap and name a few steam traps.

Blank Page

# 3. Electrical analogue steady-state conduction heat transfer

**Nomenclature**

A    Heat transfer area, m$^2$

E    Electrical potential, V

h    Heat transfer coefficient, W/m$^2$K

i    Current, Amp

k    Thermal conductivity, W/mK

L    Distance between two planes, cm

Q    Rate of heat transfer, W

r    Resistance, Ohm

$r_{th}$   Thermal resistance, K/W

T    Thermal potential

$V_P$   Potential scale factor, K/V

$V_R$   Resistance scale factor, K/W Ohm

ΔE   Potential difference, V

ΔT   Temperature difference, K

## 3.1 Introduction

Analogous are extensively used in heat transfer studies. The most useful analogous are electrical ones for linear processes. The important application is in understanding and conceptualization process. For example, the distribution of current in a net work, electrical analogue is easier to visualize than the flow of heat in a continuous conduction region. Analogue methods have been successfully used to determine the heat transfer rates in complicated geometries. Further the electrical analogue may be used in three dimensional problems.

Electrical analogue is not applicable to problems involving distributed heat generation.

## 3.2 Objective of the experiment

To solve the following problem using network type analogue.

Problem: A hollow rectangular duct has inside dimensions of 2 m x 1m, of wall thickness 1m and outer dimensions of 4m x 3m. The inner walls are at uniform temperature of 700°C and the outer walls at 100°C. Find the temperature distribution and steady heat losses per unit height by the network analogue method. Draw the isothermal and adiabatic lines. Compare the heat losses calculated by the above method with the analytical approach neglecting the end losses.

Thermal conductivity of duct material may be taken as 1.5 W/mK.

## 3.3 Theory

The process of steady state electrical conduction in a distributed resistance is directly analogous to steady state heat conduction. The differential equations relating electrical potential E and thermal potential T, to the space coordinates are

$$d^2E/dx^2 + d^2E/dy^2 + d^2E/dz^2 = 0 \qquad\qquad 3..1$$

and

$$d^2T/dx^2 + d^2T/dy^2 + d^2T/dz^2 = 0 \qquad\qquad 3.2$$

The two dimensional form is similar to equations (3.1) and (3.2) but has only two terms. These equations are of the Laplace form. The corresponding variables are E and T and any scale factor may be chosen. The analogous electrical and thermal quantities are as follows:

| Electrical | Thermal |
|---|---|
| Potential difference :$\Delta E$ | Temperature difference: $\Delta T$ |
| Resitance : r | Thermal resistance : $r_{th}$ |
| | = L/kA |
| Current : I | Heat flow : Q = $\Delta T$ / $r_{th}$ |

The above analogies can be better understood from the Fourier heat conduction equation.

$$Q = k \, A \, \Delta T / L = \Delta T / (L/kA) = \Delta T / r_{th} \qquad 3.3$$

Here the term $\Delta T$ corresponds to the temperature difference between two planes at a distance L apart.

To calculate the temperature and quantity of heat, two scale factors viz., potential scale factor $V_P$ and resistance scale factor $V_R$ are defined as

$$V_P = \Delta T / \Delta E \, V_r \qquad 3.4$$
$$V_R = r_{th} / r \qquad 3.5$$

The value of $r_{th}$ is numerically equal to $k^{-1}$ per unit length.

## 3.4 Experimental set-up

Fig 3.1 is an electrical analogue net work. Here only one fourth section of the duct of the problem in section 3.2 is considered. The thickness of the wall is divided in four equal sections. The heat transfer problem chosen here is very simple. In many cases, the area perpendicular to conduction heat transfer can not be covered completely by equal size squares. Often the bounding surfaces may not be isothermal.

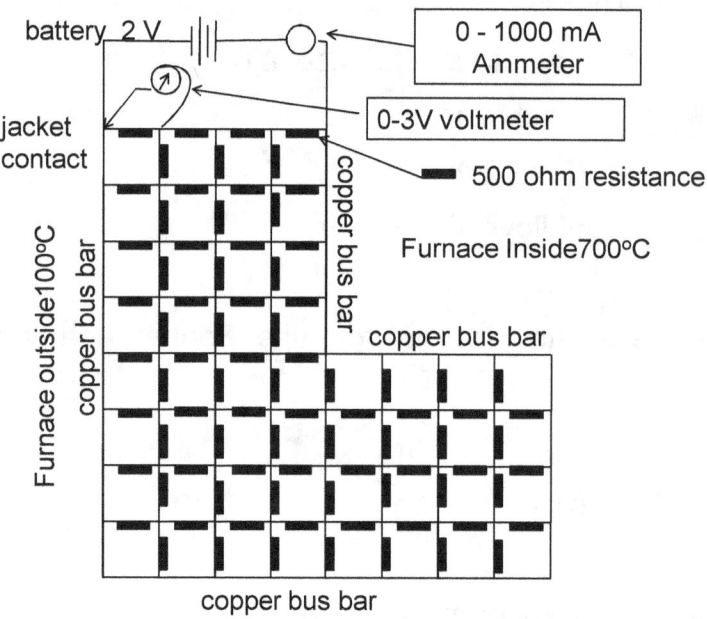

Fig. 3.1 The electrical net work analogue circuit

A 20 cm x 20 cm x 3 mm ebonite piece is taken. Resistances of same values (say around 50 ohms) are connected as shown in Fig.3.1. The side conductors (those corresponding to the isothermal surfaces) are copper plates of 1 cm by 2 mm cross section. The inner side corresponds to 700°C and outer side corresponds to 100°C

## 3.5 Experimental procedure

A fully charged lead-acid cell is connected across the bus bars as shown in Fig. 3.1. The voltage of the cell should not be less than 2 volts. A mill- ammeter with a range 0 – 1000 mA is connected in series with the net work. The intermediate voltage are measured with the help of jockey contact at all nodal points (ie., junctions of the resistances)

## 3.6 Presentation of data

A diagram similar to Fig. 3.1 is drawn on a separate sheet of paper and all the potential differences are noted.

## 3.7 Results

From the figure drawn as indicated in section 3.6, the intermediate temperatures corresponding to the nodal voltages are calculated using the voltage scale factors. Isothermal (ie., constant temperature ) lines are drawn and from them adiabatic lines (these lines are perpendicular to isothermal lines ) are drawn.

The quantity of heat per unit time per unit length of duct is given by

$$Q/L = \Delta T/r_{th} = V_P \, \Delta E/ \, VR \; r_{th} = V_P/V_R \; I \qquad\qquad 3.6$$
$$= k \, V_P \, r_i \qquad\qquad 3.7$$

The total heat flow per unit length is equal to four times the value calculated from equation (3.7)

Thus the steady state heat loss $= 4 \, k \, V_P \, r_i$           3.8

The amount of heat loss determination from the electrical net work is compared with that calculated analytically.

## 3.8 Questions

    (1) Explain why the isothermal line and adiabatic lines are at right angles?

    (2) Mention a few situations where the use of analogues is more useful?

    (3) How are electrical conductivity and thermal conductivity are related for pure metals

Blank Page

# 4. Unsteady state heat transfer

**Nomenclature**

A   Heat transfer area, $m^2$

$C_p$   Specific heat of metal, J/kgK

h   Heat transfer coefficient, $W/m^2K$

k   Thermal conductivity, W/mK

L   Height of cylinder, m

r   Distance in radial direction, m

R   Radius of the cylinder, m

T   Temperature of the cylinder at any time, $^{\circ}C$

$T_o$   Initial temperature of the surrounding medium, $^{\circ}C$

$T_{\infty}$   Unchanging temperature of the surrounding medium, $^{\circ}C$

V   Volume of the cylinder, $m^3$

Dimensional groups

$N_{Bi}$ Biot number, (k/hR)

$N_{Fo}$ Fourier number, ($\alpha\theta/R^2$)

Greek letters

$\alpha$   Thermal diffusivity, ($k/\rho_s C_p$), $m^2/s$

$\theta$   Time, s

$\rho_s$   Density of cylinder, $kg/m^3$

## 4.1 Introduction

An understanding of unsteady-state heat conduction (both transient and periodic) is very essential since it plays an important role in many heat transfer process. For example, designers in technological areas are often faced with start-up and other operating transients well to guide material selection. Unsteady conduction is involved in the quenching of billets, the annealing of solids, the manufacture of glass, the burning of bricks, the steaming of wood, and the vulcanization of rubber. When a body is heated continuously, the temperature at a given point within the body asymptotically approaches the temperature of the medium. The points near the surface quickly approach the temperature of the surroundings, and those in the interior lag far behind.

## 4.2 The objective of the experiment

To estimate the film heat transfer coefficient between the medium in which the body is heated or cooled.

## 4.3 Theory

Consider an arbitrary solid object of volume V, surface area A, density $\rho_s$ and specific heat $C_p$. It is assumed to be initially at temperature $T_0$. This object is exposed suddenly to an environment at temperature $T_\infty$, greater than $T_0$.

   At any time $\theta$, the rate of increase of energy content in the solid material is equal to the rate of heat transport from the surroundings at $T_\infty$.

The energy balance equation is given by

$$\rho \, C_p \, V \, dT/d\theta = h \, A \, (T_\infty - T) \qquad\qquad 4.1$$

$$\text{Thus} \quad dT/d\theta = (h \, A \, / \, \rho_s \, C_p \, V) \, (T_\infty - T) \qquad\qquad 4.2$$

Assuming the value of ($h \, A \, / \, \rho_s \, C_p \, V$) remains constant during the time interval $\theta$, equation (4.2) is integrated with the following initial and boundary conditions.

At $\theta = 0$, $T = T_o$ and at $\theta = \theta$, $T = T$

$$\int_{T_o}^{T} dT/(T_\infty - T) = (h\,A/\,\rho_s\,C_p\,V) \int_{0}^{\theta} d\theta \qquad\qquad 4.3$$

$$\ln\left[(T_\infty - T)/(T_\infty - T_o)\right] = (h\,A\,\theta)/(\rho_s\,C_p\,V) \qquad\qquad 4.4$$

In case of an infinitely long cylinder, the dimensionless temperature is given by

$$(T_\infty - T)/(T_\infty - T_o) = f\left(\alpha\theta/R^2,\ k/hR,\ r/R\right) \qquad\qquad 4.5$$

When the temperature is measured at the center of the cylinder, equation (4.5) simplifies as

$$(T_\infty - T)/(T_\infty - T_o) = f\left(\alpha\theta/R^2,\ k/hR\right) \qquad\qquad 4.6$$
$$= f\left(N_{Fo},\ N_{Bi}\right) \qquad\qquad 4.6a$$

## 4.4 Experimental set-up

The experimental set-up is shown in Fig.4.1. A constant temperature water bath serves as the hot environment. A brass cylinder of height 15 cm and diameter 6 cm is the solid object. This cylinder is provided at the center with a previously calibrated iron-constantan thermocouple (30 Gauge) located at half of the cylinder height. This measures the center temperature of the cylinder. The bath temperature is measured with a mercury-in-glass thermometer (0-110°C).

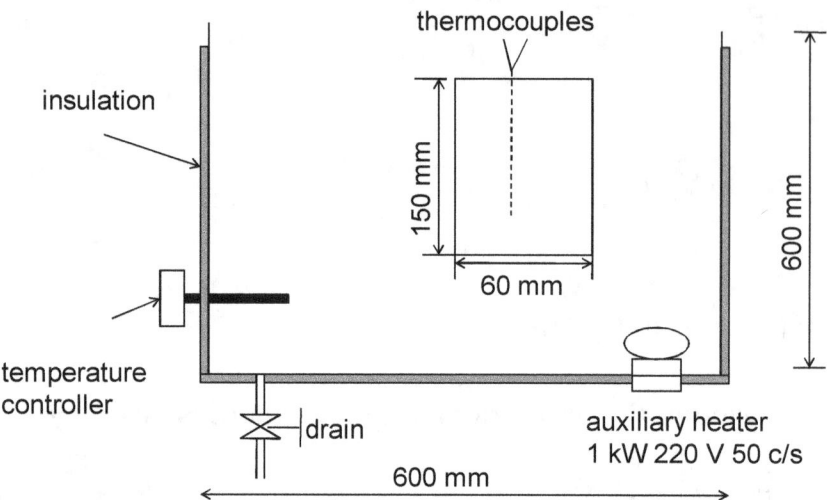

Fig. 4.1 Experimental set-up for unsteady-state
Heat transfer studies

## 4.5 Experimental procedure

The cylinder is suddenly immersed in the constant temperature bath. Immediately the thermocouple voltage and corresponding time are noted till the voltage read by millivoltmeter dies not change with time. Now the cylinder is removed from the water bath and cooling rate is also noted keeping the cylinder in ambient air.

## 4.6 Presentation of data

Measure the diameter and height of the cylinder.

Tabulate the following both for heating and cooling processes.

   (1) Time, s
   (2) Thermocouple reading, mV
   (3) Temperature of the cylinder, $^{o}C$
   (4) Ambient temperature, $^{o}C$

## 4.7 Results

A plot of log $(T_\infty - T) / (T_\infty - T_0)$ against $\alpha\theta/R^2$ is made on a semi-log paper. The slope of the line is compared with the Gurnie -Luri charts available in pages 10-8 to 10-9, Perry, Chemical Engineers Hand Book, 5 th ed. Thus the value of the parameter (kh/R) is obtained. The value of liquid film heat transfer coefficient h is calculated knowing the physical properties of the material. This method is followed to determine the air film heat transfer coefficient during the cooling period also.

Note: - Different diameters and lengths of cylinders of different physical   properties can be used. Recommended materials are copper and stainless steel

## 4.8 Questions

1) Explain periodic and aperiodic transient conduction
2) What do you mean by unsteady state process?
3) What is meant by time constant?
4) Does the film coefficient change with material of the solid?
5) Show that Fourier number is dimensionless.

Blank Page

# 5. Effective thermal conductivity of a packed bed

**Nomenclature**

| | |
|---|---|
| A | Cross sectional area of the bed, $m^2$ |
| D | Diameter of the packed bed, m |
| $D_c$ | Diameter of the copper rod, m |
| $d_p$ | Equivalent diameter of packing material, m |
| $k_c$ | Thermal conductivity of copper, W/mK |
| $k_{eff}$ | Effective thermal conductivity of packed bed, W/mK |
| $L_1$ | Bed height, m |
| $L_2$ | Distance between the points 3 and 4, m (Fig. 5.1) |
| Q | Rate of heat transfer, W |
| $T_1, T_2, T_3, T_4$ | Temperatures at points, 1, 2, 3 and 4, $^oC$ (Fig. 5.1) |
| $TC_1, TC_2, TC_3, TC_4$ | Thermocouples at points, 1, 2, 3 and 4 (Fig. 5.1) |

## 5.1 Introduction

Studies on heat transfer rates in packed beds are of the frequent use of packed catalyst beds in chemical reactors where considerable amount of thermal energy is being absorbed or released. The energy transport equations require the value of the effective thermal conductivity of the packed bed. The effective thermal conductivity is dependent on the bed voidage.

## 5.2 Objective of the experiment

To determine the effective thermal conductivity of a packed bed.

## 5.3 Theory

Consider a packed bed of height $L_1$, whose ends are kept at different temperatures $T_1$ and $T_2$. The cross sectional area of the bed is A. If $T_1 > T_2$, the steady state heat transfer through the bed is given by

$$Q = k_{eff} A (T_1 - T_2)/L \qquad\qquad 5.1$$

The equation is valid only if radial temperature distribution ( ie., dT/dr) is negligible or zero.

The effective thermal conductivity $k_{eff}$ can be calculated from the values of Q, $T_1$, $T_2$, A and L.

## 5.4 Experimental set-up

The details of the apparatus used to determine the effective thermal conductivity of a packed bed are shown in Fig. 5.1. The packed bed may be of 3 mm porcelain beads placed between two copper plates inside a silica tube. Two calibrated iron-constantan thermocouples $TC_1$ and $TC_2$ of 30 Gauge are embedded in the copper plates. Copper is preferred to other materials because of its high thermal diffusivity. The temperatures measured by these two thermocouples are the temperatures at two ends of the packed bed. Two more thermocouples $TC_3$ and $TC_4$ of the same type and gauge as $TC_1$ and $TC_2$ are embedded at a known distance $L_2$ inside the copper rod, which is brazed to the bottom copper plate of the packed bed. A heater is placed above the top copper plate to supply heat to the packed bed. The heater is made by wrapping Nichrome wire (of resistance 11 ohms) on a mica sheet cut to the diameter of the silica tube. The power is supplied through a Variac (input 230V, out put 0 – 260 V A.C 50c/s). The copper rod is cooled at its bottom end by circulating water through the chamber which helps in quicker establishment of steady state. The complete assembly is insulated to minimize heat losses.

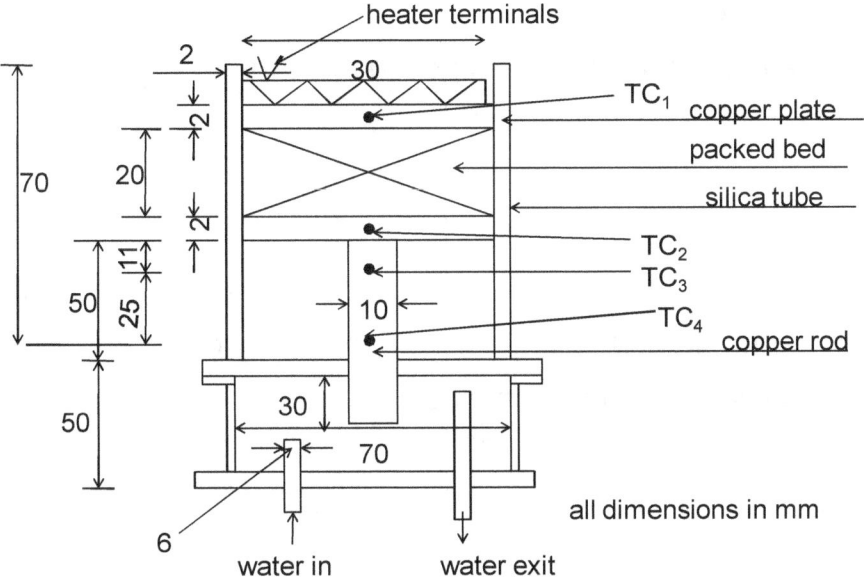

Fig. 5.1 Experimental set-up for determining the
effective thermal conductivity of a packed bed

## 5.5 Experimental procedure

The packed bed height over the bottom copper plate is noted. The heater is placed over the upper copper plate. All the thermocouple leads are connected to a Millivoltmeter through a selector switch. The heater is connected to a Variac. Cooling water circulation is started. After steady state is obtained the temperature measured by the thermocouples $TC_1$, $TC_2$, $TC_3$ and $TC_4$ are noted. The experiment is repeated for different values of heat input to the heater and different cooling water flow rates.

## 5.6 Presentation of data

Note following for different sets of data

(1) Height of the packed bed, m

(2) Diameter of the packed bed, m

(3) Thermal conductivity of copper, W/mK

(4) Diameter of the copper rod, m

(5) All thermocouple readings in mV are converted to temperatures.

## 5.7 Results

Assuming no heat loss from the surface of copper rod, the rate of heat flow through the packed bed is equated to that flowing through the copper rod across the distance $L_2$. Thus the heat balance equation (5.1) can be rewritten as

$$Q = k_c \pi D^2c\ (T_3\text{-}T_4)/(4\ L_2)$$
$$= k_{eff}\ (\pi D^2/4\ L_1)\ (T_1\text{-}T_2) \qquad\qquad 5.2$$
$$\text{or}\quad k_{eff} = (k_c\ L_1/L_2)\ (D^2{}_c/D^2)[\ (T_3\text{-}T_4)/(T_1\text{-}T_2)] \qquad\qquad 5.3$$

The results are tabulated as follows

(1) Flow rate of water/heat input

(2) Effective thermal conductivity, $k_{eff}$, W/mK

Note: - Various packing materials may be used. For example, ceramic or glass beads, Raschig rings, Lessing rings, Paul rings, Berl saddles etc,. It is essential that the ratio of the column diameter to particle diameter is more than 10 to avoid wall effects.

## 5.8 Questions

(1) How the effective thermal conductivity depends on bed voidage?

(2) Between copper and glass, which has more thermal diffusivity?

(3) Mention the use of the effective thermal conductivity.

# 6. Heat transfer by free convection

**Nomenclature**

A   Area of metal object exposed to air, $m^2$

B   Plate width, m

$C_f$   Specific heat of air, J/kgK

$C_p$  Specific heat of metal, J/kgK

g   Acceleration due to gravity, $m/s^2$

h   Film heat transfer coefficient, $W/m^2K$

$h_a$  Natural convection film heat transfer coefficient, $W/m^2K$

$k_f$   Thermal conductivity of air, W/mK

L   Length of the plate, m

T   Temperature of plate at any time $\theta$, $^{\circ}C$

$T_s$  Average plate temperature, $^{\circ}C$

$T_o$  Initial temperature of the plate, $^{\circ}C$

$T_\infty$ Ambient air temperature, $^{\circ}C$

V  Volume of the plate, $m^3$

W  Width of the plate, m


Dimensionless groups


$N_{Gr}$   Grashof number, $L^3 \rho_f^2 \, g \, \beta\Delta T/\mu_f^2$


$N_{Nu}$   Nusselt number, $h_a L / k_f$


$N_{Pr}$   Prandtl number, $C_f \, \mu_f / k_f$

Greek letters

$\Delta T$  Temperature difference, $(T-T_\infty)$, $^{\circ}K$

$\beta$     Coefficient of volume expansion, $K^{-1}$

$\rho_f$    Density of air, $kg/m^3$

$\rho_s$    Density of metal, $kg/m^3$

$\mu_f$     Viscosity of air, $Ns/m^2$

$\theta$    Time, s

## 6.1 Introduction

Consider a warm plate exposed to air at room temperature. The density of the air is very near to the plate will be less than that of the main body of the air, since the temperature of the air near the hot surface is more than in bulk. Thus buoyant forces cause an upward flow of air near the surface. Heat is transmitted through gas layers and is carried away by bulk motion or convection. Although both conduction and convection are involved, this process is called Natural or Free convection. Design of fins for cooling the automobile engine and calculation of optimum thickness of insulating material to minimize the heat losses, are examples where one requires the values of convective heat transfer coefficients.

## 6.2 Objective of the experiment

To find out the natural convective heat transfer coefficient for vertical and horizontal plates.

## 6.3 Theory

For detailed theory, refer section 4.3. The equation (4.4) is repeated here for convenience.

$$\ln [(T_\infty - T)/(T_\infty - T_0)] = (h\,A\,\theta)/(\rho_s\,Cp\,V) \qquad 4.4$$

When equation (4.4) is plotted on a semi-log paper as $[(T_\infty - T)/(T_\infty - T_0)]$ against time $\theta$, the slope of the line is $(h\,A)/(\rho_s\,Cp\,V)$. From this value the film coefficient is computed.

The correlations available in literature are summarized below for both vertical plates and for horizontal plates.

Vertical plates
For short vertical plates not more than 60 cm height, the Nusselt number $N_{Nu}$ is given by
$$N_{Nu} = 0.52\,(N_{Gr}\,N_{Pr})^4 \qquad\qquad 6.1$$

Horizontal plates
For heated plates facing upward or cooled plates facing downward for turbulent range
$2 \times 10^7 < N_{Gr}\,N_{Pr} < 3 \times 10^{10}$
$$N_{Nu} = 0.14\,(N_{Gr}\,N_{Pr})^{1/3} \qquad\qquad 6.2$$

For laminar range
$1 \times 10^5 < N_{Gr}\,N_{Pr} < 2 \times 10^7$
$$N_{Nu} = 0.54\,(N_{Gr}\,N_{Pr})^{1/4} \qquad\qquad 6.3$$

For heated plates facing downward or cooled plates facing upward,
For laminar range
$3 \times 10^5 < N_{Gr}\,N_{Pr} < 3 \times 10^{10}$
$$N_{Nu} = 0.27\,(N_{Gr}\,N_{Pr})^{1/4} \qquad\qquad 6.4$$

**6.4 Experimental set-up**

The experimental set-up is very simple and schematic is in Fig 6.1. A thin copper plate 50 mm x 20 mm x 3 mm is taken. Calibrated thermocouples (30 Gauge) are embedded, one at the center and four at each of the corners. The plate is fixed to a stand with provision to vary the plane angle of the plate to the horizontal. The entire set-up is kept in an enclosure large enough to consider that the plate is kept in an infinite medium of air.

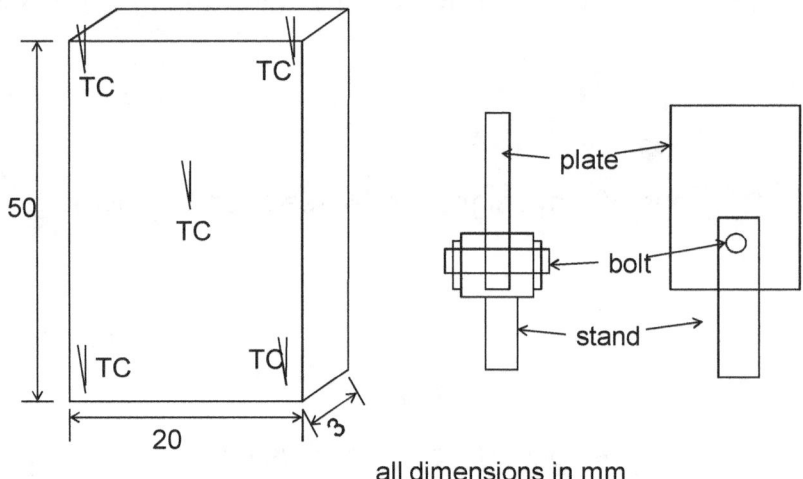

all dimensions in mm

Fig. 6.1 Experimental scheme

## 6.5 Experimental procedure

The plate is heated and all the thermocouples are connected to a mille voltmeter through a selector switch. The plate is held in vertical position and the cooling rate of the plate is noted. The average of the temperatures recorded by five thermocouples is taken as the plate temperature. The ambient air temperature is noted. The above procedure is repeated for horizontal position also. The experiment may be carried out at various inclinations.

## 6.6 Presentation of data

Note the length, width, and thickness of the plate. For each plate inclination, tabulate the following

(1) Plate position

(2) Average plate temperature, °C

(3) Time, s

(4) Ambient air temperature, $^{\circ}$C

## 6.7 Results

The values of $N_{Gr}$, $N_{Pr}$ and ($N_{Gr}$ $N_{Pr}$) are calculated to know the flow region viz. laminar or turbulent. $N_{Nu}$ is calculated using the appropriate correlation given in section (6.3) and the value of $h_a$ is computed. The experimental value of h obtained from equation (4.4) is calculated. All the values are tabulated in the following way.

(1) $N_{Gr}$

(2) $N_{Pr}$

(3) $N_{Gr}$ $N_{Pr}$

(4) $h_a$ calculated from correlations, given in section (6.3), W/m$^2$K

(5) h calculated from equation (4.4), W/m$^2$K

## 6.8 Questions

1. Explain the mechanism of Natural convection
2. Show that $N_{Gr}$ is dimensionless
3. Prove that for ideal gas the coefficient of cubical expansion is equal to inverse of the absolute temperature.

Blank Page

# 7. Double pipe heat exchanger

**Nomenclature**

$A_i$     Inside surface area of tube, $m^2$

$A_o$     Outside surface area of tube, $m^2$

$C_p$     Specific heat of water, J/kgK

$C_1$, $C_2$, $C_3$, $C_4$, $C_5$ constants appearing in text

$D_i$     Inside diameter of tube, m

$D_o$     Outside diameter of tube, m

$D_{lm}$     Logarithmic mean of inside and outside diameters,

       $(D_o - D_i)$ / Ln $(D_o/D_i)$, m

F      Geometric factor, dimensionless

$h_i$     Inside film heat transfer coefficient, $W/m^2K$

$h_o$     Outside film heat transfer coefficient, $W/m^2K$

$k_f$     Thermal conductivity of water, W/mK

$k_t$     Thermal conductivity of tube material, W/mK

L      Length of the heat transfer section, m

Q      Rate of heat transfer, W

$T_1$     Water inlet temperature, $^{\circ}C$

$T_2$     Water outlet temperature, $^{\circ}C$

$U_o$     Overall heat transfer coefficient based on outside surface area of tube,

       W/mK

$V_m$     Mean velocity of water in the tube, m/s

x      Tube wall thickness, m


Dimensionless groups


$N_{Re}$    Reynolds number,    $D_i V_m \rho_f/\mu_f$

$N_{Pr}$    Prandtl number,      $C_p \mu_f/k_f$

$N_{Nu}$    Nussult number,     $h_i D_i / k_f$

Greek letters

$\Delta T_{lm}$   Logarithmic mean temperature difference, °K

$\Delta T_m$   Effective mean temperature between the fluids, °K

$\rho_f$   Density of water, kg/m$^3$

$\mu_f$   Viscosity of water, Ns/m$^2$

$\lambda$   Latent heat of vaporization, J/kg

## 7.1 Introduction

In many heat transfer processes, heat is transferred from one fluid to another through a solid wall. For example, tubular heat exchangers are frequently used in industry where heat transfer between fluid steams is desired. The exchanger geometry, fluid properties and flow rates are the parameters that influence the rate of heat transfer. A study of these parameters forms the basis for design of heat exchangers.

## 7.2 Objectives of the experiment

(a) To determine the influence of water flow rate on the overall heat transfer coefficient.

(b) To estimate the individual film heat transfer coefficients on the steam side and water side of the tube by means of Wilson plot.

## 7.3 Theory

The rate of heat transfer to the water is given by

$$Q = W C_p (T_2 - T_1)$$
$$= U_o A_o \Delta T_{lm} \tag{7.1}$$

Where $\Delta T_{lm} = F \Delta T_m$

If steam side is assumed to be at constant temperature, F equals one. The water flow rate, water inlet and outlet temperatures and the temperature of the steam permit the evaluation of $U_o$ from equation (7.1). The individual film coefficients can be computed using Wilson's plot.

## Wilson plot

For clean tubes the overall-heat transfer coefficient $U_o$ is given by

$$1/U_o = 1/h_o + (x/k_t)(D_o/D_{lm}) + (1/h_i)(D_o/D_i) \qquad\qquad 7.2$$

For a given pressure and quality of steam, the steam side film coefficient may be assumed to be constant. The term $[(x/k_t)\ (D_o/D_i)]$ contains only the tube thickness, diameters ratio and metal conductivity. It is independent of both steam pressure and water flow rate and thus remains constant. The water side coefficient $h_i$, (for Reynolds number in the range of 10,000 to 40,000 and Prandtl numbers from 0.7 to 120) can be expressed as

$$h_i D_i/k_f = C_1\ N_{Re}^{0.8}\ N_{Pr}^{0.4} \qquad\qquad 7.3$$

Where $C_1$ is a constant

If the mean water temperature in different runs does not vary too widely, then the physical properties can be assumed to remain unchanged. Thus we can write equation (7.3) as

$$h_i = C_2\ V_m^{0.8} \qquad\qquad 7.4$$

Hence equation (7.2) can be written as

$$1/U_o = C_2 + C_4 + C_5 V_m^{-0.8}$$  7.5

A plot of $1/U_o$ vs $V_m^{-0.8}$ (the Wilson plot) should give a straight line of slope $C_5$ with interception equal to $(C_3+C_4)$ at $V_m^{-0.8} = 0$. the value of $C_4 = (x/k_t) (D_o/D_{lm})$ in equation (7.2) can be calculated and hence $C_3$ $(=1/h_o)$.

Since $C_5 = (D_o/D_i)(1/C_2) = (D_o/D_i)( V_m^{-0.8} /h_i)$, $h_i$ is calculated as a function of $V_m$ from the values of $h_o$ and $h_i$ thus calculated are to be compared with the available correlations. For convenience the correlations are presented below.

 (i) Nusselt equation for film condensation on a horizontal tube

$$h_o = 0.725 (k_f^3 l\rho_f^2 g / D_o \mu_f \Delta T)^{1/4}$$  7.6

 (ii) Dittus- Boeltier equation for highly turbulent flow

$$h_f = 0.023 (k_f/D_i)(D_i V_m \rho_f/\mu_f)^{0.8} (C_p \mu_f/k_f)^{0.4}$$  7.7

## 7.4 Experimental set-up

A schematic sketch is shown in Fig. 7.1. the heat exchanger consists of a copper tube (14 mm I.D. and 22.2 mm O.D.) kept concentrically in a steel pipe of 40 mm I.D. and 48 mm O.D.. The effective length of the heat exchanger is 1500 mm. provisions are made to measure water inlet and outlet temperatures, water flow rate, steam pressure, steam temperature and wall temperature. Water is passed through the exchanger in a closed circuit, and its temperature can be controlled by means of an overflow, with a corresponding make-up of cold water from the mains.

The tube wall temperature is measured with three iron-constantan thermocouples (30 Gauge previously calibrated- Refer Fig 7.1, 5 and 9) which are embedded at the mid-point of the wall. The arithmetic mean value of the thermocouple readings is taken as the mean temperature at the mid-point of the tube wall. From this, inside and outside surface temperatures of the tube are calculated.

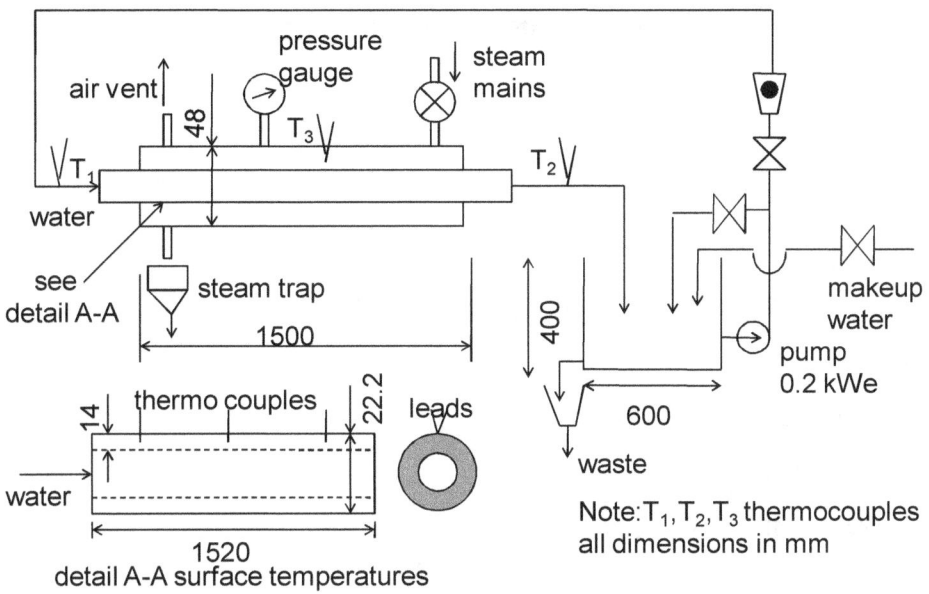

Fig. 7.1 Double pipe heat exchanger

## 7.5 Experimental procedure

The tank is filled with water to a predetermined level by make-up water. The pump is started and the water flow is set around $32 \times 10^{-6}$ kg/s by adjusting the bye-pass valve. It is very important to see that the cold side of the exchanger ie., water side is always started first to avoid damage due to thermal expansion. The steam valve is opened slowly and the pressure is kept at 150 kN/m$^2$. The water in the loop gets heated with time, so the make up water is used to maintain a constant difference of 10 to 20$^{\circ}$K between steam and water exit temperature ($T_2$). Steady state condition is said to be reached when, the over flow, make up flow, the tank level, the steam pressure and the wall temperatures remain unchanged with time. Soon after the steady state is attained, note the steam pressure, its temperature, water flow rate, water inlet and out let temperatures and tube wall temperatures. The above procedure is repeated for increasing water flow rates from $32 \times 10^{-6}$ kg/s till maximum attainable flow rate covering at least six to eight water flow rates.

Important: At the end of the experiment the steam supply should be cut-off. Then allow the water to circulate for a few minutes before stopping the pump.

## 7.6 Presentation of date

Tabulate the following

(1) Steam condensing temperature at steam pressure, $^\circ$C

(2) Water flow rate, $^\circ$C

(3) Average tube wall temperature, $^\circ$C

(4) Water inlet temperature, $^\circ$C

(5) Water out let temperature, $^\circ$C

## 7.7 Results

For each set of readings evaluate $U_o$, $h_o$ and $h_i$ from equation (7.1) and by Wilson plot method. The values of $h_o$ and $h_i$ calculated above are compared with the values of $h_o$ and $h_i$ from equations (7.6) and (7.7). For comparison tabulate the following.

(1) Run number

(2) $h_o$ from Wilso plot, W/m$^2$K

(3) $h_o$ from equation (7.6), W/m$^2$K

(4) $h_i$ from Wilson plot, W/m$^2$K

(5) $h_i$ from equation (7.7), W/m$^2$K

## 7.8 Questios

(1) Define overall heat transfer coefficient.

(2) In a double pipe heat exchanger air is heated by hot water. In this situation which side has maximum resistance heat transfer?

(3) Define log-mean temperature difference.

(4) Define temperature range and temperature approach.

Blank Page

# 8. Finned tube heat exchanger

**Nomenclature**

$A_b$  Bare area of the tube, m$^2$

$A_F$  Fin area, m$^2$

$A_i$  Inside area of tube, m$^2$

$A_o$  Outside area the tube, m$^2$

  ($\eta\, A_F + A_b$)

$C_p$  Specific heat of air at constant pressure, J/kgK

$D_e$  Equivalent diameter, m

$D_i$  Inside diameter, m

$D_{lm}$ Logarithmic mean of inside and outside diameters, m

$G_f$  Mass velocity of air, kg/m$^2$s

$h_i$  Inside film coefficient, W/m$^2$K

$h_o$  Outside film heat transfer coefficient, W/m$^2$K

$J_H$  Colburn JH factor (Dimensionless)

$k_m$  Thermal conductivity of tube material, W/mK

$k_l$  Water thermal conductivity at condensing temperature, W/mK

$L_p$  Perimeter of fin, m

m  Condensate flow rate, kg/s

P  Perimeter, m

Q  Rate of heat transfer, W

S  Cross-sectional area of the fin, m$^2$

$T_1$  Air inlet temperature, °C

$T_2$  Air out let temperature, °C

$T_3$  Steam saturation temperature corresponding to steam pressure, °C

$U_o$  Overall heat transfer coefficient based on outside area, W/m$^2$K

V  Air superficial velocity, m/s

W  Mass flow rate of air, kg/s

$x_F$  Length of fin, m

$x_W$   Tube wall thickness, m

Dimemensionless groups

$N_{Pr}$   Prandtl number, $Cp\,\mu_f/k_f$

$N_{Pr}$   Reynolds number, $D_e\,G_f/\mu_f$

$N_{Nu}$   Station, $h_o/C_p\,\rho_f\,V$

Greek letters

$\lambda$   Latent heat of vaporization, J/kg

$\eta_F$   Fin efficiency

$\alpha$   $\sqrt{(ho\,LP/(S\,k_m)}$, $m^{-1}$

$\rho_f$   Density of air at the mean temperature of $T_1$ and $T_2$, $kg/m^3$

$\rho_l$   Density of water at condensing temperature, $kg/m^3$

$\mu_f$   Viscosity of air at the mean temperature of $T_1$ and $T_2$, $Ns/m^2$

$\mu_l$   Viscosity  of water at condensing temperature, $Ns/m^2$

$\rho_g$   Density of steam at prevailing steam pressure, $kg/m^3$

$\Delta T$   $T_2$-$T_1$, $^oK$

$\Delta T_1$  $T_3$-$T_1$, $^oK$

$\Delta T_2$  $T_3$-$T_2$, $^oK$

$\Delta T_{lm}$   $(\Delta T_1-\Delta T_2)/\ln(\Delta T_1/\Delta T_2)$, $^oK$

## 8.1 Introduction

Extended surface heat exchangers are required in cases where the unit area for heat transfer on one side of the exchanger is much less than the other. Or one of the heat exchanging fluids has lower film heat transfer coefficient in comparison with the other. For instance, consider a double pipe heat exchanger in which fluid on out side of the tube is condensing steam and inside air or some other gas flowing. Normally the condensing steam film coefficients are of the order 100 to 200 times that of air. If the heat transfer surfaces on both sides of the heater are nearly equal, then the heat transfer rate is controlled by fluid that was low film coefficient. More over the overall heat transfer coefficient will not be much different from the gas film coefficient. Thus it is obvious that any improvement in the heat transfer must be by way of increasing the heat transfer area on the gas side rather than on the steam side.

The method most usually adopted is to increase the heat transfer surface area on the gas side by welding fins on the surface.

Extended surfaces are neither efficient nor necessary if the film coefficient on both sides is of the same magnitude. Further fins increase the pressure drop.

8.2 Objective of the experiment

To determine the overall heat transfer coefficients based on gas side and compare it with the value obtained from available correlations.

**8.3 Theory**

Equating the heat picked up by air stream to that transferred from inside to air side, one gets

$$Q = W\, C_p\, (T_2 - T_1) = U_o\, A_o\, \Delta T_{lm} \qquad\qquad 8.1$$

$$\text{Thus } U_o = W\, C_p\, (T_2 - T_1) \,/\, (\, A_o\, \Delta T_{lm}\, ) \qquad\qquad 8.2$$

General theory

Consider unit length of tube. Let $A_F$ be the fin area and $A_b$ bare tube area. The outside film coefficient ho is assumed to be the same for both the fins and the tube. The overall heat transfer coefficient, based on the outside area can be written as

$$1/U_o = A_o/h_i A_i + (x_w A_o)/(k_m D_{lm}) + 1/h_o \qquad 8.3$$

Where $A_o = \eta_F A_F + A_b$

To use equation (8.3) it is necessary to know the values of the fin efficiency $\eta_F$ and the individual coefficients $h_o$ and $h_i$. The coefficient $h_i$ in the present case is steam condensing coefficient in a horizontal tube and is given by

$$h_i = 0.612 \left[(k_l^3 \, \rho_l \, (\rho_l-\rho_g) \, g \, \lambda)(\mu_l D_i \Delta T)\right]^{1/4} \qquad 8.4$$

**Definition of Fin efficiency**

A unit area of the fin surface is not as effective as a unit area of bare tube surface for heat transfer. This is because of the temperature drop along the length of the fin. However if the fin material has infinite thermal conductivity, the temperature will be uniform throughout.

**Fin efficiency is defined as**

$\eta_F$ = Heat transfer rate from free surface / Heat transfer rate from an identical fin of infinite thermal conductivity

= Heat transfer rate from free surface / Heat transfer rate from fin if it is at base temperature

The efficiency of longitudinal fins is given in Fig 8.1. In this $\eta_F$ is plotted as a function of the quantity $\alpha_F \, \eta_F$. The product $\alpha_F \, \eta_F$ is dimensionless. Fig 8.1 is drawn from $\alpha_F \, \eta_F$ and $\eta_F$ values given. Table 8.1.

Table 8.1 Fin efficiency

| $\alpha_F\,\eta_F$ | 0 | 0.5 | 1.0 | 1.5 | 2.0 | 2.5 | 3.0 | 3.5 | 4.0 |
|---|---|---|---|---|---|---|---|---|---|
| $\eta_F$ | 1.0 | 0.91 | 0.75 | 0.6 | 0.5 | 0.4 | 0.31 | 0.29 | 0.28 |

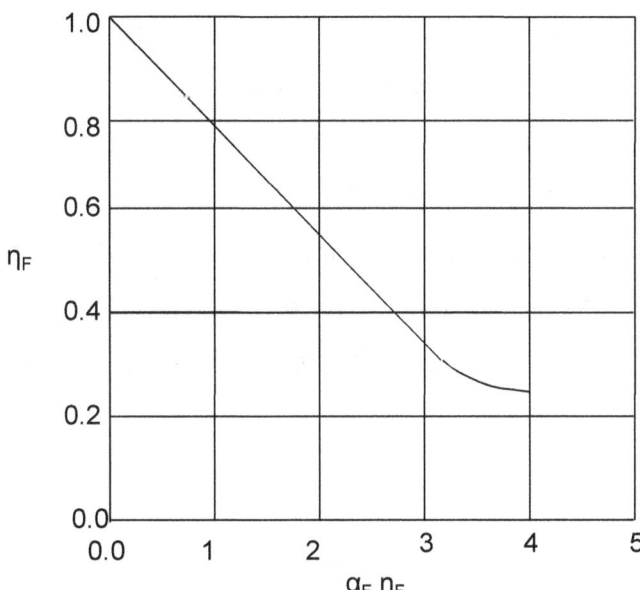

Fig. 8.1 Fin efficiency-longitudinal fin

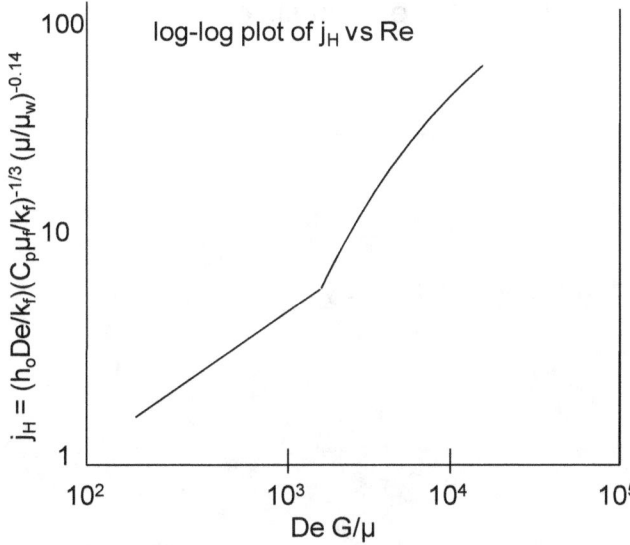

Fig. 8.3 Longitudinal fin heat transfer

The outside film coefficient ($h_o$) is calculated from the equation (8.3). The value of $h_o$ thus calculated is compared with the plot of $J_H$ vs $N_{Re}$ given in Fig. 8.3.

## 8.4 Experimental set-up

The experimental set-up is shown in Fig. 8.2. The heat exchanger consists of a 40 mm steel pipe having 25 longitudinal fins (15 mm high and 1 mm thick). The effective heat exchanger length is 1.2 mm. this tube is enveloped by 80 mm steel tube. Steam condenses in the inner pipe and air flows through the annulus over the fins. Inlet and outlet air temperatures are measured with mercury-in glass thermometers. Air flow rate is measured with an orifice meter. The condensing tube is provided with a pressure gauge, air vent and a steam trap.

## 8.5 Experimental procedure

Air is admitted to the apparatus from a blower. The flow rate of air is kept at a particular value (around $8 \times 10^{-3}$ kg/s). Steam is admitted to the inner tube and the pressure is kept around 150kN/m$^2$. After attaining steady state condition, the inlet and outlet air temperatures are noted. At the same time the steam pressure and condensate flow are noted. The experiment is repeated for two pressures and six different air flow rates up to the maximum attainable.

## 8.6 Presentation of data

The experimental data are tabulated in the following way

   (1) Steam pressure. kN/m$^2$
   (2) Manometer reading, mm WC
   (3) Inlet air temperature, $^{\circ}$C
   (4) Outlet air temperature, $^{\circ}$C
   (5) Mass flow rate of air, kg/s

(6) Condensate flow in 1 minute, kg

(7) Condensate flow rate, kg/s

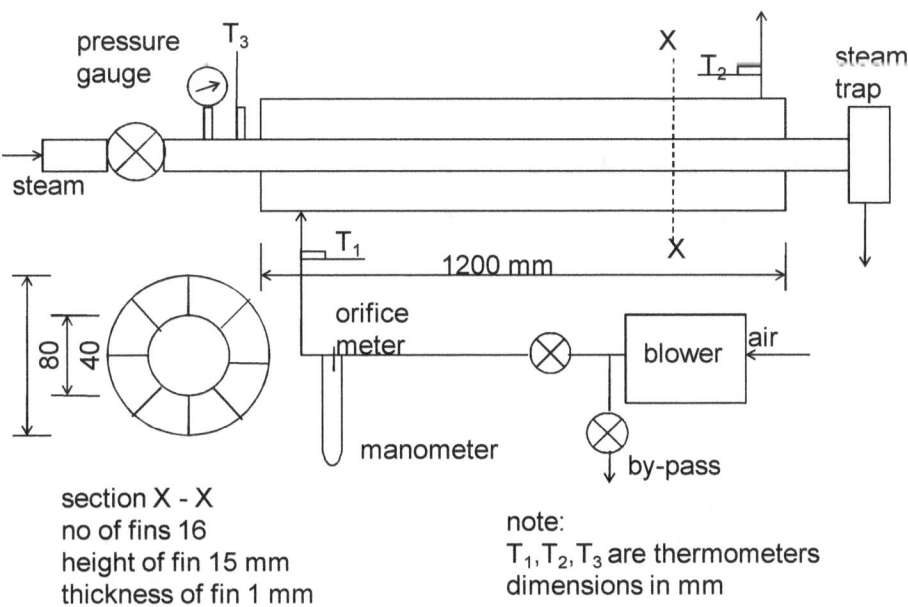

section X - X
no of fins 16
height of fin 15 mm
thickness of fin 1 mm

note:
$T_1, T_2, T_3$ are thermometers
dimensions in mm

Fig. 8.2 Finned tube heat exchanger

## 8.7 Results

First a heat balance is made between the amount of heat taken by the air and the condensate rate

$$Q = W \, C_p \, (T_2 - T_1) = m \, \lambda \qquad\qquad 8.5$$

The value of $h_o$ calculated for different air flow rates by equation (8.3) is compared with the value of $h_o$ obtained from the plot of $J_H$ vs $N_{Re}$ mentioned in section 8.3.. The results are tabulated as below

(1) Steam pressure, $kN/m^2$

(2) Air flow rate, kg/s

(3) $h_o$ obtained from equation (8.3), W/m²K

(4) $h_a$ calculated from $J_H$ vs $N_{Re}$ plot, W/m²K

## 8.8 Questions

(1) What is meant by fin efficiency?

(2) Give two examples where longitudinal fins and transverse fins are used.

(3) What is the advantage of a finned tube (or extended surface) heat exchanger?

(4) On what basis the material of construction of fin is chosen?

# 9. Shell and tube heat exchanger

## Nomenclature

$A_o$ Outside area of the tubes, $m^2$

$C_3$, $C_4$, $C_5$ Constants appearing in text

$C_p$ Specific heat of water, J/kgK

$D_2$ Outer diameter of a tube, m

g Acceleration due to gravity, $m/s^2$

$h_o$ Film heat transfer coefficient, $W/m^2K$

$k_f$ Thermal conductivity of condensate film, W/mK

L Heated tube length, m

m Mass flow rate of water, kg/s

Q Heat transfer rate, W

$T_1$ Water inlet temperature, $^oC$

$T_2$ Water outlet temperature, $^oC$

$T_3$ Steam inlet temperature, $^oC$

$T_4$ Steam outlet temperature, $^oC$

$T_f$ Condensate film temperature, $^oC$

$T_v$ Steam temperature, $^oC$

$T_w$ Tube wall temperature, $^oC$

$U_o$ Overall heat transfer coefficient based on outside area, $W/m^2K$

$V_m$ Water average velocity, m/s

w Condensate rate, kg/s

Greek letters

$\rho_f$ Density of condensate, $kg/m^3$

$\lambda$ Latent heat of vaporization, J/kg

$\mu_f$ Condensate film viscosity, $Ns/m^2$

$\Delta T$ Temperature drop across condensate film, $^oK$

$\Delta T_{lm}$ Logarithmic mean temperature difference, $^oK$

## 9.1 Introduction

When it is required to have large heat transfer area in a relatively small volume either more tubes passes or sometimes both are to be provided. Therefore double pipe heat exchanger is not used in industry since number of required becomes so large that the shell and tube construction is more economical.

## 9.2 Objective of the experiment

To determine overall heat transfer coefficient experimentally and compare it with the value calculated from available correlations.

## 9.3 Theory

When a saturated vapor is brought in contact with a cooled surface, heat is transferred from the vapor to the surface and a film of condensate is produced.

During this process one can obtain either film wise or drop wise condensation. If the surface is wettable, film wise condensation occurs: other wise drop wise condensation occurs. However one cannot completely eliminate mixed condensation, ie, the combination of film wise and drop wise condensation.

The heat transfer coefficients obtained during film wise are one fifth to one sixths of drop wise condensation. In industrial condensers, film wise condensation occurs unless promoters are added to sustain drop wise condensation.

In the case of vapor condensation on a vertical tube, condensate film flows downward under the influence of gravity, but is retarded by the viscosity of the liquid. This flow will normally be streamline and heat will flow by conduction. Nusselt developed a theoretical relation to determine the film heat transfer coefficient in terms of the physical properties and temperature differences.

The film coefficient for condensation over a vertical plate of height, L, is given by

$$h_o = 0.943 \left[ (k_f^3 \, \rho_f \, g \, \lambda) / (L \, \mu_f \, \Delta T) \right]^{0.25} \tag{9.1}$$

Film condensation over outside of a vertical tube is given by

$$h_o \, D_o / k_f = 0.0077 \left[ (k_f \, L \, \Delta T) / (D_2 \, \lambda \mu_f) \right]^{0.67} (D_2^3 \, \rho_f^2 \, g / \mu_f^2)^{0.56} \tag{9.2}$$

The physical properties are evaluated at the film temperature $T_f$

Where $T_f = [T_v - (3/4)(T_v - T_w)]$

## 9.4 Experimental set-up

A typical set-up is shown in Fig.9.1: consists of a tube bundle heat exchanger mounted vertically. It has 12 tubes, with 10 mm nominal bore. Each tube length is 600 mm and the tubes are arranged in triangular pitch of 25 mm.

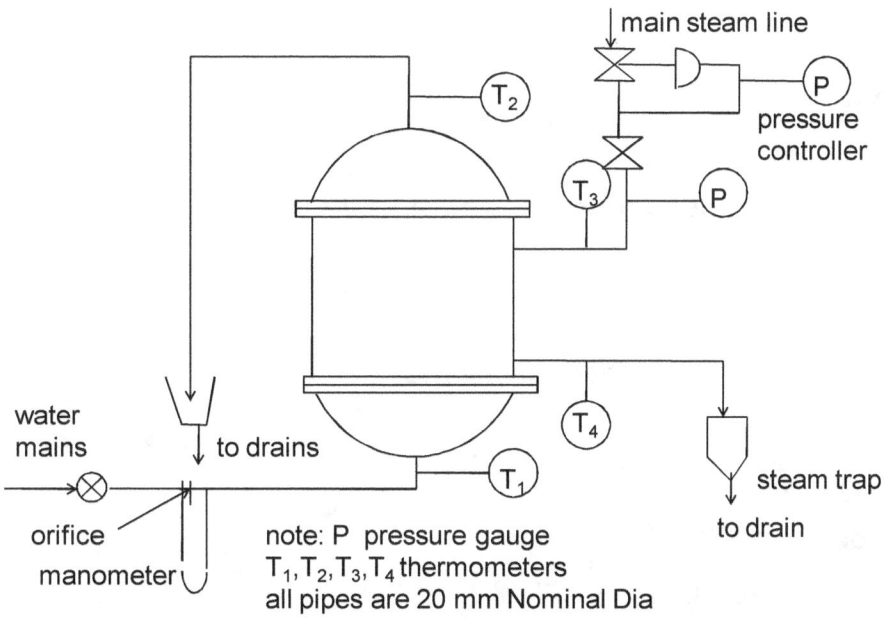

Fig. 9.1 Shell and tube heat exchanger

These are housed in a shell of 150 mm nominal bore (IS 1239). Shell and tubes are made of steel. Tube sheets are of the fixed type. Water passes through the tubes and steam condenses over the tubes. A sketch of the lay-out including all pipe lines is shown in Fig. 9.1. Steam is supplied to the unit from a boiler. The condensate is removed through a steam trap.

A pressure regulator is used to control the pressure of the incoming steam. Four thermometers are provided to measure the incoming and outgoing water, steam and condensate temperatures.

## 9.5 Experimental procedure

Before admitting the steam to the shell side of the condenser, the cold water should be passed through the tubes. The shell side pressure is maintained around 150 kN/m$^2$. When the temperature of the leaving water reaches steady state value, the steam pressure, steam inlet and outlet temperatures, condensate rate and water flow rate, inlet and outlet water temperatures are noted. This experiment is repeated for various liquid velocities and for two steam pressures. The liquid flow rate is varied in such a way that turbulent flow always persists inside the tubes.

## 9.6 Presentation of data

Note the number of tubes, tube length, inside and outside diameters and shell inside diameter.

Experimental observations may be tabulated as follows

(1) Steam pressure, kN/m$^2$
(2) Steam inlet temperature, $^\circ$C
(3) Steam outlet temperature. $^\circ$C
(4) Water flow rate, kg/s
(5) Water inlet temperature, $^\circ$C
(6) Water outlet temperature, $^\circ$C
(7) Condensate collected in 2min., kg
(8) Condensate rate, kg/s

## 9.7 Results

The heat balance between the amount of heat given to water by steam and the amount of heat actually taken by the water is made from the equation (9.3)

$$Q = W\, C_p\, (T_1 - T_2) = m\, \lambda \qquad\qquad 9.3$$

The overall heat transfer coefficient is calculated from the equation (9.4)

$$U_o = m\, C_p\, (T_1 - T_2) \,/\, (A_o\, \Delta T_{lm}) \qquad\qquad 9.4$$

The individual heat transfer coefficients are determined from the Wilson's method (given in section 7.3). The value of $h_o$ obtained from the above method is compared with the value of $h_o$ from equation (9.)

For ready comparison the calculated values of $h_o$ may be tabulated as follows.

(1) Steam pressure, $kN/m^2$

(2) Condensing film coefficient $h_o$ calculated from Wilson method, $W/m^2K$

(3) Condensing film coefficient $h_o$ coefficient from equation(9.2) $W/m^2K$

## 9.8 Questions

(1) Define film wise and drop wise condensation

(2) Normally condensers are provided with an air vent; explain the necessity of air vent.

(3) How does pressure of inert gas in steam affect the condenser performance?

(4) Draw a typical plot of temperature of tube side fluid and condensing steam as a function of tube length for a tubular condenser.

Blank Page

# 10. Heat transfer in agitated vessels

**Nomenclature**

$A_i$   Inside surface area of the coil, m$^2$

$A_o$   Outside surface area of the coil, m$^2$

$A_w$  Mean of inside and outside areas of the coil, m$^2$

$C_p$  Specific heat of water, J/kgK

d     Inside diameter of a straight pipe, m

$d_c$   Diameter of helix, m

$d_i$    Inside diameter of coil, m

$d_o$   Outside diameter of coil, m

$d_v$   Internal diameter of vessel, m

$d_w$  Mean diameter of pipe wall, m

$h_i$    Inside film coefficient, W/m$^2$K

$h_o$   Outside film coefficient, W/m$^2$K

k     Thermal conductivity of water, W/mK

L     Length of paddle, m

m     Mass flow rate of water, kg/s

N     Number of revolutions per unit time, s$^{-1}$

$T_1$   Water inlet temperature, °C

$T_2$   Water outlet temperature, °C

$T_3$   Bath temperature, °C

$U_s$  Overall heat transfer coefficient based on outside area of coil, W/m$^2$K

u     Average velocity of water in the coil, m/s

$x_w$  Coil wall thickness, m

Greek letters

ρ   Density of water, kg/m$^3$

μ   Viscosity of water corresponding to bulk liquid temperature, Ns/m$^2$

$μ_s$  Viscosity of water corresponding to wall temperature, Ns/m$^2$

$\Delta T \qquad T_2 - T_1, \, ^{\circ}K$

$\Delta T_1 \qquad T_3 - T_1, \, ^{\circ}K$

$\Delta T_2 \qquad T_3 - T_2, \, ^{\circ}K$

$\Delta T_{lm} \quad (\Delta T_1 - \Delta T_2)/ \ln(\Delta T_1/\Delta T_2), \, ^{\circ}K$

## 10.1 Introduction

In most industrial chemical reactions, certain quantity of heat is to be added or removed to control the rate of reaction. The addition or removal of heat is done by passing steam or water through a jacket fitted to the outside of the vessel or through a helical coil fitted in the vessel. In both the cases bath liquid is agitated to obtain even distribution of heat in the vessel.

## 10.2 Objective of the experiment

To find out the overall heat transfer coefficient based on the outside area, for various liquid velocities and degree of agitation.

## 10.3 Theory

Thermal resistance to heat transfer arises due to water film on the inside of the coil, the tube wall, the water film on the outside of the coil and the scale that may be formed on either side of the tube surface. The overall heat transfer coefficient based on the coil outside area can be written for smooth tubes as

$$1/U_o \;\; = A_o/h_i \, A_i + (x_w/ \, K_w)(A_o/A_w) + 1/h_o \qquad\qquad 10.1$$

or

$$1/U_o \;\; = d_o/h_i \, A_i + (x_w/ \, K_w)(d_o/d_w) + 1/h_o \qquad\qquad 10.2$$

For straight pipes the inside film coefficient for Reynolds number more than 10000 is given by

$$h_i d_i / k_i = 0.023 \, (du\rho/\mu)^{0.8} \, (C_p\mu/k)^{0.4} \qquad\qquad 10.3$$

The film coefficient for liquids flowing in coils is given as

$$h_i(\text{coil}) = h_i \,(\text{straight pipe}) \, (1 + 3.5 \, d_i / d_e) \qquad\qquad 10.4$$

The value of the outside film coefficient $h_o$, for circular tanks using paddle type of agitator is given by

$$(h_o \, d_v/k)(\mu_s/\mu)^{0.14} = 0.87 \, (C_p\mu/k)^{1/3} \, (L^3 N\rho/\mu)^{0.62} \qquad\qquad 10.5$$

The overall heat transfer coefficient can be calculated from experimental data using the relation

$$U_o = (mC_p \, \Delta T) /(A_o \, \Delta T_{lm}) \qquad\qquad 10.6$$

## 10.4 Experimental set-up

A typical vessel is shown in Fig 10.1. It consists of a steam jacketed vessel

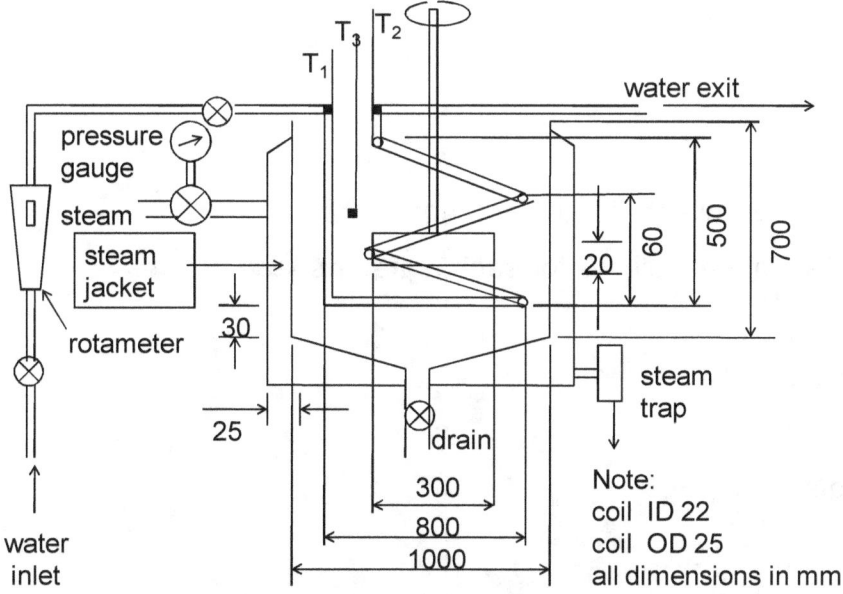

Fig. 10.1 Agitated vessel

whose inside diameter is 1 m with a wall thickness of 3 mm. The annulus of steam jacket is about 25 mm wide. The vessel is provided with a paddle or propeller type of agitator of 30 cm in diameter and its speed can be varied from 0 to 1000 rpm. The vessel is provided with a copper cooling coil 25 mm outside diameter and 22 mm inside diameter wound in the form of a helix, 80 cm in diameter. The gap between the turns in the coil is 60 mm. The clearance between bottom of the vessel and the bottom turn of the coil is 30 mm. The depth of the tank is 70 cm. the coil height is 50 cm. The speed of the stirrer is measured with a tachometer.

## 10.5 Experimental set-up

The vessel is filled with water to such a level that both the jacket portion and the helix are below the water level. Steam at 1,5 bar is then admitted to the jacketed portion. After the bath liquid is brought to its saturation temperature, water is passed through the coil at a constant flow rate. At steady state ( as noted by constant outlet temperature) the inlet and outlet water temperature, and water flow rate are noted. Six different water flow rates in turbulent flow are used. The same procedure is repeated for various agitator speeds.

## 10.6 Presentation of data

Note the coil dimensions, effective heat transfer length
Tabulate the experimental data as follows for each agitator speed:
    (1) Tank temperature, $^{o}$C
    (2) Water inlet temperature, $^{o}$C
    (3) Water outlet temperature, $^{o}$C
    (4) Water flow rate, kg/s

## 10.7 Results

The values of overall heat transfer coefficients calculated for each water flow rate and for constant speed of agitation by equations (102, 10.3, 10.4 and 10.5) are compared with the value determined from equation (10.6)

All the experimental data and calculated values may be tabulated as follows for a particular agitator speed.

(1) water flow rate, kg/s

(2) $h_i$ value from equation (10.4) , $W/m^2K$

(3) $h_o$ value from equation (10.5) , $W/m^2K$

(4) $U_o$ value from equation (10.2) , $W/m^2K$

(5) $U_o$ value from equation (10.6) , $W/m^2K$

Plots are made between Uo and agitator speed for different liquid flow rates.

## 10.8 Questions

(1) Explain Dean's effect.

(2) Distinguish between free and convective heat transfer.

(3) At zero stirrer speed, what are the dimensionless groups those exit in equation (10.5)?

(4) Distinguish between tube and pipe.

Blank Page

# 11. Heat transfer to Boiling liquids

Nomenclature

A   Heat transfer area, $m^2$

C   Specific heat, J/kgK

D   Diameter of wire, m

h   Boiling heat transfer coefficient, $W/m^2K$

g   Acceleration due to gravity, $m/s^2$

$g_c$   Gravitational conversion factor, dimensional less

I    Current, Amp

k   Thermal conductivity, W/mK

L   Length of the wire, m

Q   Heat transfer rate, W

q   Heat flux, $W/m^2$

R   Wire resistance, Ohm

T   Temperature, $^oC$

V   Voltage, V


Greek letters

$\lambda$    Latent heat of vaporization, J/kg

$\mu$    Viscosity, $Ns/m^2$

$\gamma$    Surface tension, N/m

$\rho$    Density, $kg/m^3$

$\Delta T$   $(T_w - T_L)$, $^oK$

$\Delta T_s$   $(T_w - T_s)$, $^oK$


Subscripts

L   Liquid

S   Saturation

V   Vapor

W   Wire surface


## 11.1 Introduction

The phenomena of boiling occur in many unit operations like, evaporators, distillation, steam generators etc. During boiling, a greater heat flux is transferred from a hot surface to the adjacent boiling liquid compared with non-boiling situation, for the same temperature differences between the bulk liquid and heating surface.

## 11.2 Objective of the experiment

To determine pool boiling heat transfer coefficients during sub-cooled and saturated boiling till burn out, for different wall and liquid temperature differences.

## 11.3 Theory

When a mass of liquid is below its saturation temperature and the heater surface is at a temperature above the liquid boiling point, sub-cooled (surface) boiling occurs. On the other hand if the mass of liquid is at its boiling temperature, saturated boiling occurs. For sub-cooled boiling a flow system is required and the difference between the saturation temperature and bulk liquid temperature is called sub-cooling.

Consider the boiling of water on an electrically heated horizontal platinum wire. Fig. 11.1 is a plot of heat flux against the wire and liquid temperature differences. In the range AB, liquid is superheated very near the wire and heat transfer from wire to the bulk liquid occurs by natural convection. In this region, heat flux q, is proportional to $(\Delta T)^{5/4}$. In the range BC, bubbles originate at active nuclei on the heater surface, detach and rise through the liquid pool, setting up circulation currents. This region is called nucleate boiling region, and q is proportional to $(\Delta T)^{n}$, where n ranges from 3 to 4.

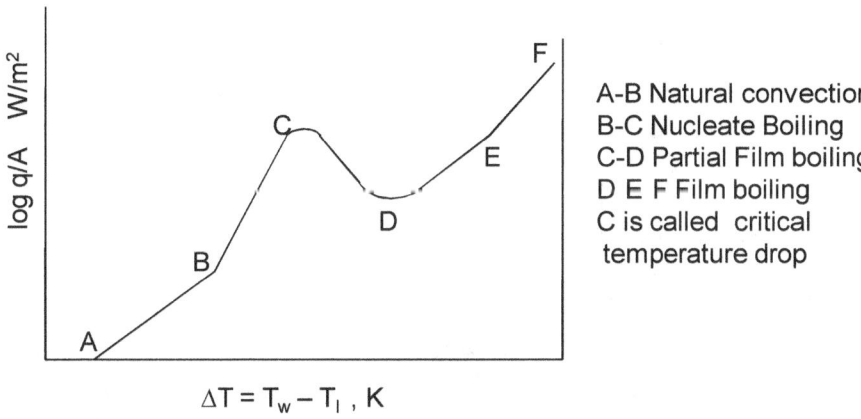

Fig 11.1 Boiling curve

At point C, the heat flux goes through a maximum or peak heat flux, also called the critical heat flux. Usually the point C is called burn-out point and the corresponding ($\Delta T$) is known as critical temperature difference. In the range CD, the surface starts getting insulated with vapor and q decreases even though ($\Delta T$) increases. This region is called transition boiling and the flux goes to a minimum at point D, called Leiden frost point. In film boiling region DEF, heat is transmitted through the vapor film by conduction and radiation. The point F corresponds to the melting temperature of the wire.

For saturated pool boiling of water the heat flux is given by the correlation

$$C_L (\Delta T)_s/\lambda = 0.013 (Q/A)/(\mu_L\lambda)\sqrt{[(g_c \gamma)/(g (\rho_L - \rho_V))]}1/3 (C_L\mu_L/k_L) \qquad 11.1$$

Where

$$Q = V I$$
$$A = \pi D L$$

## 11.4 Experimental set-up

The experimental set-up is shown in Fig 11.2. The boiling apparatus consists of a steel tank 20 cm in diameter and 40 cm. height. The tank is insulated outside to minimize heat losses to the surroundings. The tank is provided with an auxiliary heater (1 kW) and a drain. Actual boiling studies are made with a platinum wire of 15 cm in length and 0.25

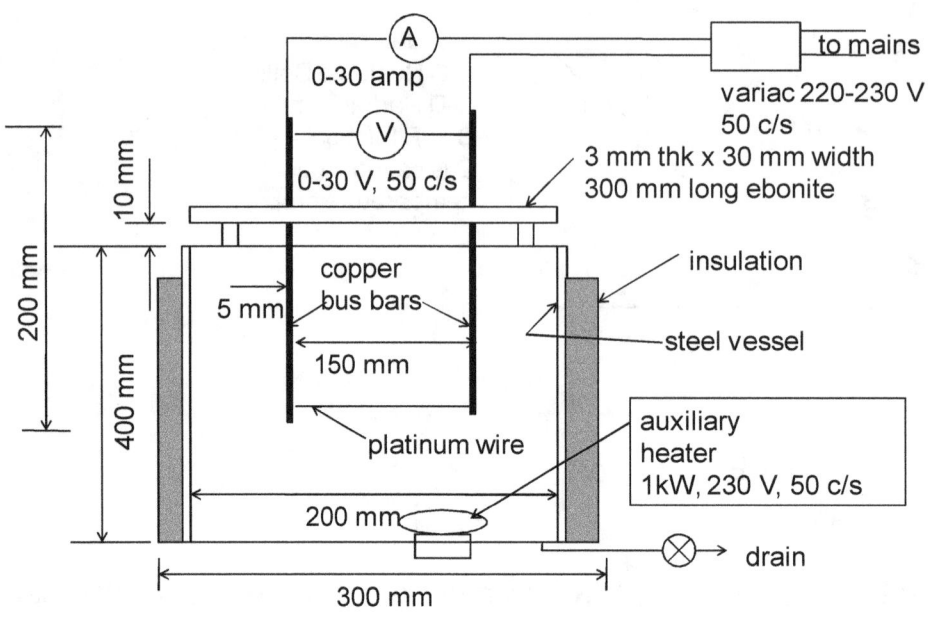

Fig 11.2 Pool boiling apparatus

mm diameter. The ends of the wire are soldered to two copper bus-bars (5 mm in diameter and 20 cm in length) as shown in Fig. 11.2. These two pass through an ebonite strip (3 mm thick, 3 cm wide and 30 cm long). Electrical connections are made as given in Fig. 11.2.

## 11.5 Experimental procedure

<u>Sub-cooled boiling</u>

The tank is filled with water. The platinum wire with its assembly is dipped in water and the electrical connections are made. Both voltage and current readings are noted in steps of 1 amp, till a maximum current of 30 amperes. The same is repeated in decreasing order also. Between each reading a time interval of one minute is allowed for steady state to establish. The liquid temperature is noted with mercury-in-glass thermometer at the beginning and at the end of the experiment. An average of these two is taken as the bulk liquid average temperature.

<u>Saturated boiling</u>

During the experiment the bulk liquid is kept at its boiling temperature by switching on the auxiliary heater. Now electricity is passed through the platinum wire noting both voltage and current as done in sub-cooled boiling. The wire surface temperature during the experiment is to be calculated from its specific resistance and thermal coefficient of resistance.

## 11.6 Presentation of data

Note the length and diameter of the wire. The experimental data is tabulated as follows for both surface boiling and saturation boiling situations.

(1) Voltage, V
(2) Current, A
(3) Resistance of wire, Ohm
(4) Wire surface temperature, $^\circ$C
(5) Bath average temperature, $^\circ$C

## 11.7 Results

The boiling heat transfer coefficient h  is given by

$$h = V I /(\pi DL(\Delta T)) \hspace{4cm} 11.2$$

The results are tabulated as below both for saturation and sub-cooled boiling.

(1) Heat transfer rate, W
(2) Heat flux, W/m$^2$
(3) Temperature difference, $^\circ$K

Plots of heat flux and heat transfer coefficient against $\Delta T$ are made for both cases.

In the case of boiling of saturation liquid the heat flux experimentally obtained is compared with the value obtained from equation (11.1). A separate plot is drawn as $(Q/A)_{Exp}$, and $(Q/A)_{Eq(11.1)}$ against $(\Delta T)$ s.

## 11.8 Questions

(1) Distinguish between boiling and evaporation

(2) What are the liquid physical properties that contribute to boiling heat transfer coefficients?

(3) Explain the role of critical temperature difference in the design of evaporators or boilers.

(4) Explain the mechanism of boiling.

# 12. Heat transfer to gas fluidized beds

**Nomenclature**

A     Heat transfer area, $m^2$

B     Cross-sectional area of the bed, $m^2$

$C_p$    Specific heat of gas at constant pressure, J/kgK

$C_s$    Specific heat of solid, J/kgK

D     Inside diameter of the column, m

$d_p$    Particle diameter, m

G     Mass velocity of gas, $kg/m^2s$

h     Heat transfer coefficient, $W/m^2K$

I     Current, Amp

k     Thermal conductivity of gas, W/mK

L     Length of bed, m

Q     Heat transfer rate, W

$Q_f$    Volumetric flow rate of air, $m^3/s$

$T_{avg}$ Average bed temperature, $^oC$

$T_w$    Heater wall temperature, $^oC$

V     Voltage, Volt

v     Superficial gas velocity, m/s


Greek letters

$\varepsilon$    Bed voidage, Dimensionaless

$\rho$    Density of gas, $kg/m^3$

$\rho_s$    Density of solid, $kg/m^3$

$\mu$    Viscosity of gas, $Ns/m^2$

$\Delta T$    $(T_w - T_{avg})$, $^oK$

$\Delta V$   Voltage difference as given by a Millivoltmeter, mV


Dimensionless groups

$N_{Rep}$   Reynolds number   $(d_p G/\mu)$

## 12.1 Introduction

Fluidized beds are used extensively in industry because of their ability to rapidly transport heat and maintain uniform temperature with in the bed. One of the potential uses of fluidized bed is for "quenching" or " freezing" hot reacting mixtures to obtain useful reaction intermediates, for example, acetylene and ethylene from hydrocarbon feed stocks.  Presently fluidization techniques are adopted in the design of fluid bed combustor in fluid bed boilers

## 12.2 Objective of the experiment

To find out the wall heat transfer coefficients for various air velocities both in fixed and fluidized beds and to compare the coefficients thus obtained with the available correlations.

## 12.3 Theory

The high magnitude of heat transfer coefficients in fluidized beds is mainly because of the break up of the laminar sub layer at the wall due to the striking action of the particles which would otherwise exist.

The heat transfer rate from a hot surface to a fixed bed or a fluidized bed is given by

$$Q = h \, A \, \Delta T \qquad\qquad\qquad 12.1$$

The heat transfer coefficient h is greater in the case of fluidized bed compared with a fixed bed for the same gas flow rate. The heat transfer coefficient depends on the physical properties and superficial velocity of gas and also on the physical properties and dimensions of the particles. The correlations for heat transfer coefficient of fixed and fluidized beds are given by equation (12.2) and (12.3) respectively as,

$$(h/C_pG)(C_p\mu/k)^{2/3} = 1.06 \ (d_pG/\mu)^{-0.41} \qquad\qquad 12.2$$

$$\text{for} \quad 350 < N_{Rep} < 4000$$

$$(hD/k) \ = 0.55 \ (D/L)^{0.65}(D/d_p)^{0.17}[(1-\varepsilon) \ \rho_s \ C_s/(\varepsilon\rho C_p)]^{0.25} \ (DG/\mu)^{0.8} \qquad 12.3$$

## 12.4 Experimental set-up

The construction details of the experimental set-up are shown in Fig. 12.1. It consists of a Perspex tube $R_1$ of 10 cm in internal diameter fixed to tube $R_2$. In the tube $R_2$ a sieve with opening of 0.5 mm is fixed. Over this sieve, a bed of glass beads 1mm diameter is packed to a height of 10 cm. The tube $R_1$ contains the guiding $H_1$ and $H_2$ through which is copper capillary tune N runs.

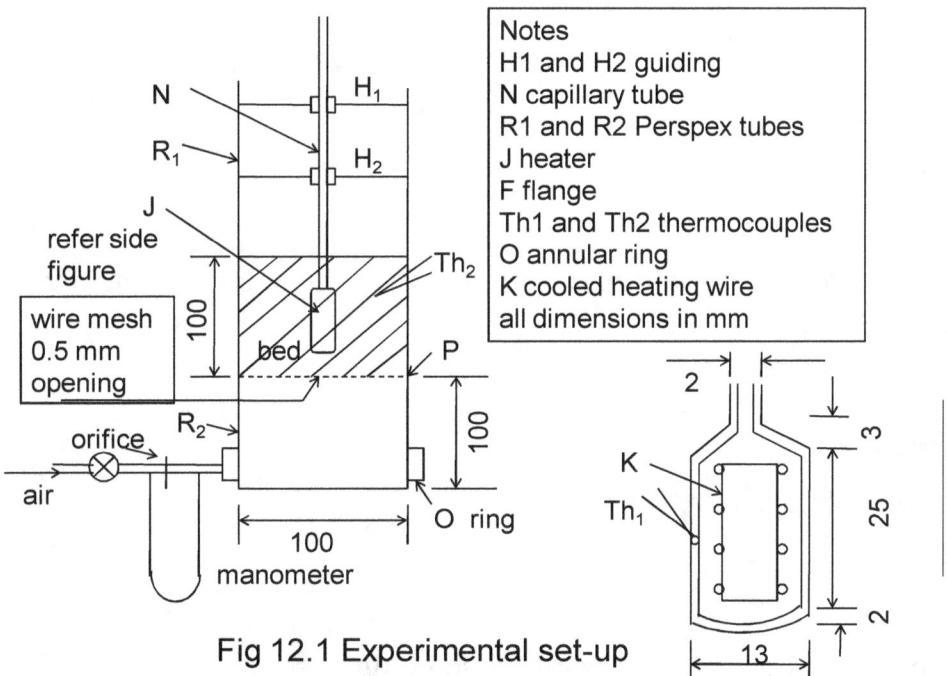

Fig 12.1 Experimental set-up

At the end of this, the heater J is fixed. The heater is made of copper with 1mm wall thickness (details are shown in Fig 12.1) and inside of this the coiled heater K is provided (resistance 140 ohms). On the surface of the heater an iron-constantan thermocouple (30 gauge) $Th_1$ is soldered. The leads for the thermocouple and the heater are passed through the capillary tube. The cold junction of this thermocouple $Th_1$ is placed in the bed half way between the heater and wall of the tube ($Th_2$). The heater is connected to power supply. Air is supplied through the ring O. The flow rate is measured with an orifice meter.

## 12.5 Experimental procedure

Air is supplied to the bed and about electric potential of 1 volt is applied to the heater. After steady state is reached, voltage, current and thermocouple readings are noted. The voltage is increased in steps of 0.5 volts till a maximum heat load of 3 watts is reached. For each voltage step, current, voltage and thermocouple readings are noted. For different gas velocities (three in fixed bed and three in fluidized bed) the readings are noted. In order to avoid fluidization of particles, that is to maintain fixed conditions, a wire mesh is kept over the bed. Thus a comparison can be made between fixed bed and fluidized bed heat transfer coefficients for any particular air velocity.

## 12.6 Presentation of data

Note the heater dimensions viz. length and diameter, tube diameter, bed height, particle diameter, fixed bed voidage.

    Tabulate the following

    (1) Volumetric flow rate of air, $m^3/s$

    (2) Voltage, V

    (3) Current, A

    (4) Thermocouples $Th_1$ and $Th_2$ voltage difference, mV

    (5) Bed voidage, $\varepsilon$

    (6) Heat transfer rate, W

    (7) Air superficial velocity, m/s

## 12.7 Results

At each air flow rate the heat transfer coefficient obtained from equation (12.1) is compared with that calculated from equation (12.2) or (12.3) as the case may be

The results are tabulated as follows:

(1) Heat transfer rate, W
(2) Heat transfer coefficient h, from equation 12.1, $W/m^2K$
(3) Heat transfer coefficient h, from equation 12.2, $W/m^2K$
(4) Heat transfer coefficient h, from equation 12.3, $W/m^2K$

At any particular heat put, plots of heat transfer coefficient against air velocity are drawn for both fixed and fluidized beds, to appreciate the advantage of fluidized beds over fixed beds.

## 12.8 Questions

(1) Explain bed voidage.
(2) Explain the wall-particle heat transfer mechanism in fluidized beds.
(3) Explain minimum fluidization velocity.
(4) Explain the advantages and disadvantages of fluidized beds and fixed beds.

Blank Page

# 13. Long tube vertical evaporator

## Nomenclature

$A$   Heat transfer area, $m^2$

$C_{pf}$ Specific heat of thin liquor, J/kgK

$C_p$   Specific heat of cooling water, J/kgK

$D$   Mean diameter of the evaporator tube, m

$L$   Effective heat transfer length, m

$m$   Mass flow rate of cooling water, kg/s

$Q$   Rate of heat transfer from the heating surface to liquid, W

$T_1$ Cooling water inlet temperature, °C

$T_2$ Cooling water outlet temperature, °C

$T$   Temperature of thick liquor, °C

$T_c$ Temperature of condensed vapor, °C

$T_f$ Temperature of thin liquor (feed), °C

$T_s$ Condensing temperature of steam, °C

$U$ Overall heat transfer coefficient, $W/m^2K$

$W$ Mass rate of thick liquor, kg/s

$W_f$ Mass rate of feed, kg/s

$W_s$ Condensate rate, kg/s

$X$ Concentration of thick liquor, mass fraction

$X_f$ Concentration of thin liquor, mass fraction

Greek letters

$\lambda$ Latent heat of condensation of steam, J/kg

## 13.1 Introduction

The process of evaporation deals with concentration of a non-volatile solute from a solution by the removal of the requisite amount of the volatile solvent. In most cases the solvent is water. Evaporation is conducted by vaporizing a portion of the solvent to produce a concentrated solution or thick liquor. Normally, in evaporation the thick liquor is the valuable product and the vapor is condensed and discarded but in desalination the vapor condensate is a valuable product.

Long tube vertical (LVT) evaporators are especially effective in concentrating liquids that tend to foam. Foam is broken when the high velocity vapor liquid mixture impinges against a vapor head baffle. Normally the fraction of liquid evaporated per pass is considerably higher in LTV evaporators than in forced circulation evaporators. The coefficients obtained in LTV evaporators are less, but the expense of operation of circulation pump is avoided.

## 13.2 Objective of the experiment

(1) To concentrate a 5 percent (by weight) sodium carbonate solution to 20 percent (by weight)
(2) To calculate the following during steady state operation
   (a) Material and heat balances
   (b) The capacity and economy
   (c) The overall heat transfer coefficients

## 13.3 Theory

**Material balance**

$$W_f \ X_f = W \ X \qquad\qquad 13.1$$

**Heat balance**

In the absence of heat losses to the surroundings and negligible heat of dilution, the heat balance equation may be written as

$$Q = W_s \lambda_s = (W_f - W) (T - T_c) + W_f C_{pf} (T - T_f) + m C_p (T_2 - T_1) \qquad \text{13.2}$$

## Capacity

Capacity may be defined as the number of kilograms of water vaporized per hour.

## Economy

Economy is defined as the number of kilograms of water evaporated per kilogram of steam used.

The heat transfer rate in an evaporator, like that of any heat transfer apparatus, is given by

$$Q = W_s \lambda_s = U A (T_s - T) \qquad \text{13.3}$$

There are two factors that affect the overall heat transfer coefficient in evaporator's viz. the boiling point elevation and hydrostatic head in the tubes.

Further the steam side film coefficient depends on the condensing temperature and the amount of non-condensed gas present in the steam. The liquid film coefficient is affected by the liquid velocity, its viscosity and cleanliness of the heating surface.

## 13.4 Experimental set-up

The experimental evaporator, shown in Fig 13.1 consists of a 20 mm (Nominal Bore) steel tube of length 2 m surrounded by a mild steel jacket 60 mm in I.D. In the annulus, steam condenses. A water level indicator shows the level of liquid in the tube. The vapor from the tube goes to a steel spherical chamber 'A' 500 mm in diameter. From the top of this, vapors are lead to a condenser. The system can be operated under vacuum. The chamber C and D are of the same size as chamber A. The condenser B is 1 meter long and 300 mm in diameter.

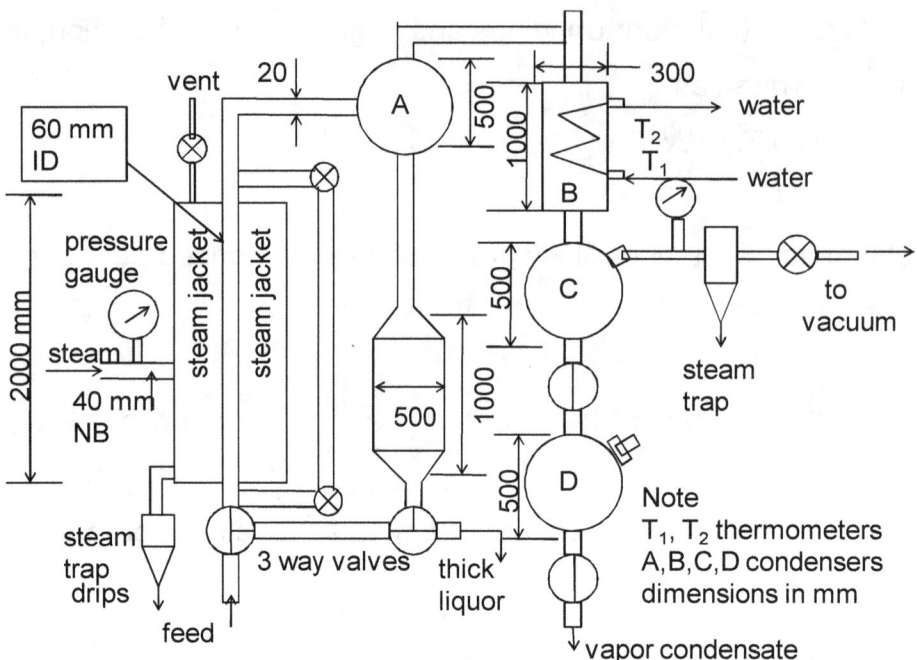

Fig 13.1 Experimental set-up

## 13.5 Experimental procedure

The experiment is done under atmospheric condition. Before admitting steam into the steam chest, 5%, (by weight) sodium carbonate solution is charged till the liquid in the tube is half full. The steam in the chest is kept around 150 kN/m². As soon as evaporation starts the liquid level in the boiler section slowly decreases. Fresh feed is admitted slowly to make up the decrease in level. The vapor condenses in the chamber B and is collected in the vessel D through C. The cold water flow rate is kept constant in the condenser coil. After the liquid is concentrated to 20%, the feed is supplied continuously such that the thick liquor concentration is maintained at 20 percent.

## 13.6 Presentation of data

Note the effective heating length and diameter of the tube
At steady state, the following are tabulated for a particular steam pressure
    (1) Condensate rate, kg/h

(2) Thin liquor flow rate, kg/s

(3) Thin liquor temperature, °C

(4) Thick liquor flow rate, kg/s

(5) Thick liquor temperature, °C

(6) Steam condensing temperature, °C

(7) Cooling water flow rate, kg/s

(8) Cooling water inlet temperature, °C

(9) Cooling water out let temperature, °C

(10) Thin liquor concentration, mass fraction

(11) Thick liquor concentration, mass fractionn

(12) Evaporator pressure, $kN/m^2$

(13) Evaporation rate, kg/s

## 13.7 Results

The results are tabulated as follows for each steam pressure

(1) Capacity

(2) Economy

(3) Overall heat transfer coefficients

## 13.8 Questions

(1) What is Duhring's rule

(2) Explain scale resistance

(3) For a triple effect evaporator battery draw a line diagram for forward feed, backward feed and mixed feed operation.

(4) What is entrainment?

Blank Page

# 14. Radiation constant

**Nomenclature**

A   Surface area, $m^2$

$C_p$ Specific heat of fluid at constant pressure, J/kgK

$C_s$ Specific heat of solid, J/kgK

g   Acceleration due to gravity, $m/s^2$

h   Heat transfer coefficient, W/ $m^2$K

$k_f$ Thermal conductivity of fluid, W/mK

L   Length, m

m   Mass of solid, kg

Q   Heat transfer rate, W

T   Temperature, °C

$T_\infty$ Ambient temperature, °C

Dimensionless groups

$N_{Nu}$  Nusselt number, $hL/k_f$

$N_{Gr}$  Grashof number, $L^3 \rho_f g \beta\Delta T/\mu_f^2$

$N_{Pr}$  Prandtl number, $C_p \mu/k_f$

Geek letters

$\varepsilon$   Emissivity

$\sigma$   Stefan-Boltzmann constant, $5.67 \times 10^{-6}$, $W/m^2K^4$

$\Omega$   Radiation constant, $W/m^2 K^4$

$\rho$   Density of fluid, $kg/m^3$

$\beta$   Coefficient of volume expansion of fluid, $K^{-1}$

$\alpha$   Absorptivity,

$\mu_f$ Viscosity of fluid, $Ns/m^2$

$\theta$   Time, s

$\Delta T$  $(T_s-T)$, K

Subscripts

1      Body 1

2      Body 2

1→2   From body 1 to 2

2→1   From body 2 to 1

s      Surface

## 14.1 Introduction

The phenomena of conduction and convection are affected by temperature differences and very little by temperature levels, where radiation heat transfer increases rapidly with temperature level. A hot body emits radiant energy in all directions, in the form of electromagnetic waves. The energy that falls on a second body is partially absorbed, partially reflected and partially transmitted. All substances above absolute zero emit radiation. Radiation due to temperature levels is called thermal radiation. This type of radiation is prevalent in kilns, crude petroleum heaters and furnaces.

## 14.2 Objective of the experiment

To determine radiation constant and emissivity of a given polished brass cylinder.

## 14.3 Theory

For a body at temperature $T_1$ enclosed by surroundings at temperature $T_2$, the rate of heat transfer by radiation from 1 to 2 is given by

$$Q_{1 \to 2} = A \, \sigma \, (\varepsilon_1 \, T_1^4 - \alpha_{1 \to 2} \, T_2^4) \qquad\qquad 14.1$$

A body which absorbs the same fraction of radiation falling on it irrespective of the wave length of the incident radiation is called a gray body. For a gray body absorptivity and emissivity are independent of temperature. Thus for gray bodies $\varepsilon_1$ may be written instead of $\alpha_{1\rightarrow2}$ in equation (14.1). Thus

$$Q_{1\rightarrow2} = 5.67\, A\, \varepsilon_1 \left[(T_1/100)^4 - (T_2/100)^4\right] \qquad\qquad 14.2$$

The radiation constant $\Omega$ is defined as

$$\Omega = 5.67\, \varepsilon 1 \qquad\qquad 14.3$$

In actual situations, the body loses heat to both by radiation and natural convection.

The convective heat transfer coefficient is correlated as

$$N_{Nu} = b\, (N_{Gr}\, N_{Pr})^n \qquad\qquad 14.4$$

The fluid properties are evaluated at the arithmetic mean temperature of the surface and the bulk fluid.

The values of b and n are given in Table 14.1 for various values of $(N_{Gr}\, N_{Pr})$.

Table 14.1

Values of b and n in Eq (14.4)

| Range of | Vertical surfaces | | Horizontal cylinders | | Horizontal plane | |
|---|---|---|---|---|---|---|
| $N_{Gr}\ N_{Pr}$ | L= height for cylinders | | L= Dia. Of cylinder or horizontal plate length | | Cooled face up ward or heated face down ward | |
| | L= Length of plate | | Heated face upward or cooled face downward | | | |
| | b | n | b | n | b | N |
| $<10^9$ | 0.13 | 0.333 | 0.13 | 0.333 | | |
| $10^4$–$10^9$ | 0.59 | 0.25 | 0.53 | 0.25 | | |
| $>10^9$ | 1.36 | 0.125 | 1.09 | 1.09 | | |
| $3\times10^5$ - $3\times10^{10}$ | | | | | 0.27 | 0.25 |

The total heat transfer between 1 and 2 may be written as

$$Q = Ai\ \Omega\ [(T_s/100)^4 - (T_{avg}/100)^4] + \sum_i h_i\ A_i\ (T_s - T_{avg}) \qquad 14.5$$

Where $A_i$ are the areas which make up the body.

The heat loss by the cylinder is

$$Q = m\ C_s\ dT/d\theta \qquad 14.6$$

## 14.4 Experimental set-up

The typical experimental set-up consists of stainless steel cylinder (80 mm in diameter and 120 mm in height). Four thermocouples are embedded evenly on the surface. The arithmetic average of the temperature indicated by the thermocouples is taken as the mean surface temperature. No figure is given.

## 14.5 Experimental procedure

The surface of the cylinder is polished well with emery paper. The cylinder (with embedded thermocouples) is heated to around 650°C in a muffle furnace. The cylinder is removed out and cooling rate is noted till the body reaches around 120°C. It is preferred to cool the body in a closed room to avoid forced convection effects.

## 14.6 Presentation of data

Note the mass, diameter and length of the cylinder and surrounding temperature $T_\infty$ .

Tabulate the following

(1) Time, s

(2) Thermocouple readings (1,2,3 and 4), mV

(3) Average surface temperature, °C

## 14.7 Results

Calculate the convective heat transfer coefficient by equation (14.4). From equations (14.5) and (14.6) determine the value of $\Omega$. Tabulate the results as below:

(1)  $\sum_i h_i A_i (T_s - T_{avg})$, W

(2)  Radiation heat transfer, W

(3)  Radiation constant $\Omega$, W/m$^2$K$^4$

(4)  Emissivity, $\varepsilon$

Note: Instead of stainless steel, brass, or iron may be used.

## 14.8 Questions

(1) What is kirchoff's law?

(2) Define 'black body' and 'gray body'.

(3) Explain view factors.

(4) What is monochromatic radiation?

Blank Page

# C++ Source Files of Experiments

## 1.Determination of thermal conductivity

```
# include <stdio.h>
# include <math.h>
# include <conio.h>
# include <stdlib.h>

 void display(void);
 void input(void);
 void output(void);
 void calc(void);
 void write(void);

 FILE *fp1;
 FILE *fp2;


 float wat,wsec,cond,consec,wcp,vap,od,km,lnth,tout,tin,mv0,mv100,dist;
 float wkgh,ckgh,dtdx,qwat,ca,kcal,q,qcnd,tc6,tc7,mvc,pi;
 float mvc1,mvc2,mvc3,mvc4,mvc5,mvc6,mvc7,tc1,tc2,tc3,tc4,tc5;

 int main(void)
 {
 display();
 input();
 calc();
 output();
 return 0;
 }

 void display(void)
 {
  window(10,10,100,50);
 }

 void input(void)
 {
 textcolor(GREEN);
 textbackground(BLACK);
 fp1=fopen("Exp1.txt","w");
 printf("1.0 DETERMINATION OF THERMAL CONDUCTIVITY \n");
 printf("DATA INPUT\n");
 cprintf("Water quaity, kg(wat) :");
 cscanf ("%f",&wat);
```

```c
cprintf("\nTime taken for collection, min (wsec):");
cscanf ("%f",&wsec);
cprintf("\nCondensate quaity, kg(cond) :");
cscanf ("%f",&cond);
cprintf("\nTime taken for collection, min (consec):");
cscanf ("%f",&consec);
cprintf("\nWater sp.heat, kJ/kg K,(wcp) :");
cscanf ("%f",&wcp);
cprintf("\nLatent heat of vaporization, kcal/kg, (vap):");
cscanf ("%f",&vap);
cprintf("\nDiameter of copper rod, mm, (od) :");
cscanf ("%f",&od);
cprintf("\nThermal conductivity of copper, kcal/hmK (km):");
cscanf ("%f",&km);
cprintf("\nLength of copper rod, mm (lnth) :");
cscanf ("%f",&lnth);
cprintf("\nOutlet water temperature, C (tout):");
cscanf ("%f",&tout);
cprintf("\nInlet water temperatueWater, C,(tin) :");
cscanf ("%f",&tin);
cprintf("\nThermo couples calibration");
cprintf("0 milli volts at 0 deg C ,(mv0) :");
cscanf ("%f",&mv0);
cprintf("\nmillivolts at 100 deg C, (mv100):");
cscanf ("%f",&mv100);
cprintf("\nDistance between two thermo couples, cm,(dist) :");
cscanf ("%f",&dist);

cprintf("\nThermocouple from steam chest sideTC1,mV (mvc1):");
cscanf ("%f",&mvc1);
cprintf("\nThermocouple TC2,mV (mvc2):");
cscanf ("%f",&mvc2);
cprintf("\nThermocouple TC3,mV (mvc3):");
cscanf ("%f",&mvc3);
cprintf("\nThermocouple TC4,mV (mvc4):");
cscanf ("%f",&mvc4);
cprintf("\nThermocouple TC5,mV (mvc5):");
cscanf ("%f",&mvc5);
cprintf("\nThermocouple TC6,mV (mvc6):");
cscanf ("%f",&mvc6);
cprintf("\nThermocouple TC7,mV (mvc7):");
cscanf ("%f",&mvc7);
fclose (fp1);
}

/* calculations*/
```

```
void calc (void)
{
pi=3.151592;
mvc=100./mv100;
tc1=mvc*mvc1;    /* Thermocouple mv to temperatures*/
tc2=mvc*mvc2;
tc3=mvc*mvc3;
tc4=mvc*mvc4;
tc5=mvc*mvc5;
tc6=mvc*mvc6;
tc7=mvc*mvc7;

wkgh=wat * 3600.0/wsec ;
ckgh=cond*3600.0/consec ;
dtdx=(tc6- tc7)/(dist/100.0);
qwat=wkgh *(tout-tin)*wcp;
ca=pi * (od/1000.)*(od/1000.0)/4.;
kcal=qwat/(ca*dtdx);
q=qwat/ca;
qcnd=ckgh*vap;
km=q/dtdx;
}

void output(void)
{
 fp2=fopen("Exp1.txt","w");
 printf("\n   INPUT");

 printf("Water quaity, kg ----------------------%8.3f\n",wat);

 printf("Time taken for collection, min-----------%8.3f\n",wsec);

 printf("Condensate quaity, kg-------------------%8.3f\n",cond);

 printf("Time taken for collection, min-----------%8.3f\n",consec);

 printf("Water sp.heat, kJ/kg K,------------------%8.3f\n",wcp);

 printf("Latent heat of vaporization, kcal/kg,----%8.3f\n",vap);

 printf("Diameter of copper rod, mm---------------%8.3f\n",od);

 printf("Thermal conductivity of copper, kcal/hmK-%8.3f\n",km);

 printf("Length of copper rod, mm-----------------%8.3f\n",lnth);
 printf("Outlet water temperature, C--------------%8.3f\n",tout);
```

```
printf("Inlet water temperatueWater, C-----------%8.3f\n",tin);
printf("\nThermo couples calibration");
printf("0 milli volts at 0 deg C ,(mv0)----------%8.3f\n",mv0);
printf("millivolts at 100 deg C-----------------%8.3f\n",mv100);
printf("Distance between two thermo couples, cm--%8.3f\n",dist);
printf("Thermocouple from steam chest sideTC1,mV-%8.3f\n",mvc1);
printf("Thermocouple TC2,mV----------------------%8.3f\n",mvc2);
printf("Thermocouple TC3,mV-------------------- %8.3f\n",mvc3);
printf("Thermocouple TC4,mV---------------------%8.3f\n",mvc4);
printf("Thermocouple TC5,mV---------------------%8.3f\n",mvc5);
printf("Thermocouple TC6,mV---------------------%8.3f\n",mvc6);
printf("Thermocouple TC7,mV---------------------%8.3f\n",mvc7);

 printf("\n RESULTS");
 printf("Water flow, kg/h,------------------ %8.3f\n",wkgh);
 printf("Condensate flow kg/h----------------%8.3f\n",ckgh);
 printf("Rate of heat transfer, kcal/h-------%8.3f\n",qwat);
 printf("Heat flux, kcal/hm2-----------------%8.3f\n",q);
 printf("dT/dx value, Deg C/m----------------%8.3f\n",dtdx);
 printf("m Cp (Tout - Tin), kcal/h-----------%8.3f\n",qwat);
 printf("Heat in condensate, kcal/h----------%8.3f\n",qcnd);
 printf("Metal therm cond (exptl),kcal/hmC---%8.3f\n",km);
 printf("TC 1 Temperature, Deg C, %8.3f\n",tc1);
 printf("TC 2 Temperature, Deg C, %8.3f\n",tc2);
 printf("TC 3 Temperature, Deg C, %8.3f\n",tc3);
 printf("TC 4 Temperature, Deg C, %8.3f\n",tc4);
 printf("TC 5 Temperature, Deg C, %8.3f\n",tc5);
 printf("TC 6 Temperature, Deg C, %8.3f\n",tc6);
 printf("TC 7 Temperature, Deg C, %8.3f\n",tc7);

fclose(fp2);
}
```

## 2.0 Insulation thickness

```
# include <stdio.h>
# include <math.h>
# include <conio.h>
# include <stdlib.h>
# include <iostream.h>

void display(void);
void input(void);
void calc(void);
void output(void);

FILE *fp1;
FILE *fp2;

float cond,tim,id,od,lngth,pr,vap,d3,r3,amb,ts,mv0,mv100;
float tc1,tc2,tc3,tc4,tc5,tc6,pi;
float tw1,tw2,kins, kair, muair, cpair, spv, tmean, beta, ro;
float ho, nu, ha, p,gr;
float qout, qout1,q, nus,n,b; ;
float tp1,tp2,mvc,t1,t2,g;
float cont,ui,r1,r2;
float hr,tw,twr,tir,uio,sp,gropr;


int main()
{
display();
input();
calc();
output();
return 0;
}

void display()
{
window(10,10,100,80);
}

void input()
{
textcolor(GREEN);
textbackground(BLACK);
fp1=fopen("Exp2.txt","w");
printf ("2.0 CRITICAL INSULATION THICKNESS\n");
```

```
printf ("DATA INPUT                          BARE\n");
cprintf("\nCondensate collected,kg              :");
cscanf ("%f",&cond);
cprintf("\nTime taken for collection, min       :");
cscanf ("%f",&tim);
cprintf("\nPipe ID, mm                          :");
cscanf ("%f",&id);
cprintf("\nPipe OD, mm                          :");
cscanf ("%f",&od);
cprintf("\nPipe length, m                       :");
cscanf ("%f",&lngth);
cprintf("\nSteam pressure, kN/m2                :");
cscanf ("%f",&p);
cprintf("\nLatent heat of vaporisation,kcal/kg :");
cscanf ("%f",&vap);
cprintf("\nOuter diameter of insulation, mm    :");
cscanf ("%f",&d3);
cprintf("\nRoom temperature, C                  :");
cscanf ("%f",&amb);
cprintf("\nCondensing steam temperature, C     :");
cscanf ("%f",&ts);
cprintf("\nThermal conduc.of insu.kcal/hm C    :");
cscanf ("%f",&kins);
cprintf("\nThermal conduc.of air, kcal/hm C    :");
cscanf ("%f",&kair);
cprintf("\nViscosity of air, kg/mh              :");
cscanf ("%f",&muair);
cprintf("\nSpecific heat of air, kcal/kg K     :");
cscanf ("%f",&cpair);
cprintf("\nmilli volts at 0 deg C ,(mv0)       :");
cscanf ("%f",&mv0);
cprintf("\nmillivolts at 100 deg C, (mv100)    :");
cscanf ("%f",&mv100);
printf("Thermo couple readings   \n");
cprintf("\nThermocouple TC1,mV                  :");
cscanf ("%f",&tc1);
cprintf("\nThermocouple TC2,mV (mvc2)           :");
cscanf ("%f",&tc2);
cprintf("\nThermocouple TC3,mV (mvc3)           :");
cscanf ("%f",&tc3);
cprintf("\nThermocouple TC4,mV (mvc4)           :");
cscanf ("%f",&tc4);
cprintf("\nThermocouple TC5,mV (mvc5)           :");
cscanf ("%f",&tc5);
cprintf("\nThermocouple TC6,mV (mvc6)           :");
cscanf ("%f",&tc6);
```

```c
printf ("NOTE:-For uninsulated pipe d3=OD\n");
fclose (fp1);
}

/* calculations */
/* cond=0.042;
 tim=10.0;
 id=19.4;
 od=25.4;
 lngth=2.0;
 p=150.0;
 vap=642.5;
 d3=25.4;
 amb=26.0;
 ts=109.0;
 kins=0.125;
 kair=0.0226;
 muair=0.0689;
 cpair=0.24;
 mv0=0.0;
 mv100=2.5;
 tc1=2.5;
 tc2=2.5;
 tc3=2.5;
 tc4=2.5;
 tc5=2.5;
 tc6=2.5; */

void calc (void)
{
pi=3.151592;
tp1 = (tc1 + tc2 + tc3 + tc4 + tc5 + tc6) / 6.;
mvc = 100.0 / (mv100 - mv0);
tw1 = mvc * tp1;

/* pipe 1 that is bare tube  */

tmean = (tw1 + amb) / 2.;

beta = 1. / (tmean + 273.);
spv = (tmean + 273.) * 359. / (273. * 28.4);
sp = 1 / spv;
ro = (sp / 62.4) * 1000.;

/* Prandtl and Grashoof numbers  */
```

```
g=9.8;
pr = cpair * muair / kair;
nu = (muair / 3600.) / ro;
gr = g * beta * (tmean - amb) * pow((d3 / 1000),3) /pow(nu,2);
pr = cpair * muair / kair;
gropr = gr * pr;

cont =cond * 60.0/tim;
q=cont *vap;

if (gropr > 1000000000.)
{
b = 0.13  ;
n = 0.333 ;
nus = b * pow(gropr,n);
}

if (gropr > 10000. || gropr <= 1000000000.)
{
b = 0.53  ;
n = 0.25  ;
nus = b * pow(gropr,n);
}

if (gropr <= 10000.)
{
b = 0.13 ;
n = 0.333 ;
nus = b * pow(gropr,n);
}

ha = nus * kair / (d3 / 1000.);

twr = (tmean * 1.8 + 32.) + 460.;
tir = (amb * 1.8 + 32.) + 460.;
hr = (0.171e-8 * 0.8 * (pow(twr,4) - pow(tir,4)) / (twr - tir)) * 4.88;
ho = ha + hr;

ui = q/ (2. * pi * (id / 2000.) * lngth * (ts - amb));
r1 = id / 2.;
r2 = od / 2. ;
r3=d3/2.0;
uio =1./ ((r1 / r2) * (1. / ho));
qout = uio * (2. * pi * (id / 2000.) * lngth * (ts - amb));

}
```

```
void output (void)
{
textcolor(GREEN);
textbackground(BLACK);
fp2=fopen("Exp2.txt","r");
printf("\n        RESULTS");
printf("\nAverage wall temperature, C,---------%8.3f\n",tw1);
printf("Condensate flow kg/h,-------------------%8.3f\n",cont);
printf("Rate of heat transfer, kcal/h----------%8.3f\n",q);
printf("Heat loss to surroundings, kcal/h------%8.3f\n",qout);
printf("1. Ui  eq 2.11, kcal/hm2 C-------------%8.3f\n",ui);
printf("2. ha eq 2.5 kcal/hm2 C----------------%8.3f\n",ha);
printf("3. hr eq 2.7 kcal/hm2 C----------------%8.3f\n",hr);
printf("4. ho eq 2.4 kcal/m2 h C---------------%8.3f\n",ho);
printf("5. ui eq 2.9 kcal/hm2 C----------------%8.3f\n",uio);
printf("Prndtl number------------------------- %8.3f\n",pr);
printf("Grashoff number----------------------- %8.3f\n",gr);
printf("Density of air, kg/m3,---------------- %8.5f\n",ro);
printf("Air cubical expansion,---------------- %8.5f\n",beta);
printf("---------------------------------------------");
printf("\n Repeat the calculations for all pipes" );
fclose(fp2);
}


/* SUB ROUTINES  */

/*  Groshoff number routine
void grpr (void)
{
g = 9.8;
nu = (muair / 3600.) / ro;
gr = g * beta * (tmean - amb) * pow((od / 1000),3) /pow(nu,2);
pr = cpair * muair / kair;
gropr = gr * pr;
}        */

/*  Nusselt number  */
/* void hout(void)
{
nus = b * pow(grpr,n);
}    */

/*  Radiation heat transfer coefficient  */
/* void radh(void)
```

```
{
float tw,tinf,hr,twr, tir ;
twr = (tw * 1.8 + 32.) + 460.;
tir = (tinf * 1.8 + 32.) + 460.;
hr = (0.171e-8 * 0.8 * (pow(twr,4) - pow(tir,4)) / (twr - tir)) * 4.88;
} */


/*  Inside heat transfercoefficient pipe 1 */
/* float pipe1()
{
double r1, ui,r2,ho;
r1 = id / 2. ;
ui =1.0/ ((r1 / r2) * (1. / ho));
return (ui);
} */


/*  Inside heat transfercoefficient pipe 2 */
/* float pipe2()
{
double r1,ui,r3,r2,ho;
r1 = id / 2.;
ui =1./ (r1 / kins * log(r3/ r2) + (r1 / r2) * (1. / ho));
return (ui);
}*/


/*  Inside heat transfercoefficient pipe 3 */
/* double pipe3()
{
double r1,ui,kins,r3,r2,ho;
r1 = id / 2.;
ui =1./ (r1 / kins * log(r3 / r2) + (r1 / r2) * (1. / ho));
return(ui);
}*/


/*  Inside heat transfercoefficient pipe 4 */
/* float pipe4()
{
double r1,ui,kins,r3,r2,ho;
r1 = id / 2.;
ui =1./ (r1 / kins * log(r3 / r2) + (r1 / r2) * (1. / ho));
return (ui);
}*/


/*  Inside heat transfercoefficient pipe 5 */
/* float pipe5()
{
```

```
double r1,ui,kins,r3,r2,ho;
r1 = id / 2.;
ui =1./(r1 / kins * log(r3 / r2) + (r1 / r2) * (1. / ho));
return (ui);
}*/
```

Blank Page

# 3.Electrical analogue

```c
# include <stdio.h>
# include <math.h>
# include <conio.h>

 FILE *fp;


 float volt,dt,ohm,amp,kins,ql,rth,vp;

 int main(void)
 {
 window(10,10,100,30);
 textcolor(GREEN);
 textbackground(BLACK);
 fp=fopen("Exp3.txt","w");


 printf("3.0 ELECTRICAL ANALOUGUEY----------- \n");

 printf("DATA INPUT\n");
 cprintf("potentail difference, volt (volt)-----");
 cscanf ("%f",&volt);
 cprintf("\ntemperature difference, C (dt)------");
 cscanf ("%f",&dt);
 cprintf("\nresistance (ohm)-------------------");
 cscanf ("%f",&ohm);
 cprintf("\ncurrent flow (amp)-----------------");
 cscanf ("%f",&amp);
 cprintf("\nDuct thermal cond.(kcal/h,C)--------");
 cscanf ("%f",&kins);

/* calculations*/

 vp=dt/volt;
 rth=1.0/kins;
 ql=4.0*kins*vp*amp*ohm;

 printf("\n          RESULTS");
 printf("\nTotal heat loss per unit length, kcal/h, %8.5f\n",ql);

 fclose(fp);
 return 0;
 }
```

Blank Page

# 4.Unsteady state heat transfer

```
# include <stdio.h>
# include <math.h>
# include <conio.h>

 FILE *fp;


  float ht,km,cp,ti,amb,mvc,t,teta,v,alpha,t40,dt,fo,pi,ro,r;


  int main(void)
  {
  window(10,10,100,30);
  textcolor(GREEN);
  textbackground(BLACK);

  fp=fopen("Exp3.txt","w");

  printf("4.0 UNSTEADY STATE HEAT TRANSFER---------");
  printf("DATA INPUT\n");
  cprintf("Cylinder height, cm--------------------");
  cscanf ("%f",&ht);
  cprintf("\nCylinder radius, cm-------------------");
  cscanf ("%f",&r);
  cprintf("\nThermal cond. of material, kcal/hm C--");
  cscanf ("%f",&km);
  cprintf("\nDensity of cylinder, kg/m3------------");
  cscanf ("%f",&ro);
  cprintf("\nSp. ht. of the material, kcak/kg C----");
  cscanf ("%f",&cp);
  cprintf("\nInitial temp. of cylinder, C----------");
  cscanf ("%f",&ti);
  cprintf("\nSurrounding temperature, C------------");
  cscanf ("%f",&amb);
  cprintf("\nFor Iron-Constantan TC, mv/40 C-------");
  cscanf ("%f",&mvc);
  cprintf("\nTC reading, mV-----------------------");
  cscanf ("%f",&t);
  cprintf("\nTime , min---------------------------");
  cscanf ("%f",&teta);


/* calculations*/
```

```c
pi=3.141592;
v=pi* r * r * ht;
alpha=km/(ro*cp);
fo=(alpha*teta/60.)/((r/100.)*(r/100.));
t40=t*mvc;
dt=(amb-t40)/(amb-ti);


printf("\n          RESULTS");
printf("\n Values required for plotting the graph");
printf("\n(Too - T)/(Too- To)---- %8.5f\n",dt);
printf("\nFourier number,-------- %8.5f\n",fo);

fclose(fp);
return 0;
}
```

# 5.Effective thermal conductivity of a packed bed

```c
# include <stdio.h>
# include <math.h>
# include <conio.h>

 FILE *fp;


 float bdh,diab,diac,dp,km,dist,mvc,pi,tc1,tc2,tc3,tc4,t1,t2,t3,t4;
 float a,q,keff;


 int main(void)
 {
 window(10,10,100,30);
 textcolor(GREEN);
 textbackground(BLACK);

 fp=fopen("Exp3.txt","w");


 printf("5.0 EFFECTIVE THERMAL CONDUCTIVITY OF A PACKED BED----- \n");

 printf("DATA INPUT\n");
 cprintf("Height of packed bed, mm------------------");
 cscanf ("%f",&bdh);
 cprintf("\nDiameter of packed bed, mm--------------");
 cscanf ("%f",&diab);
 cprintf("\nDiameter of copper rod, mm--------------");
 cscanf ("%f",&diac);
 cprintf("\nParticle diameter, mm-------------------");
 cscanf ("%f",&dp);
 cprintf("\nThermal cond. of copper, kcak/kg C-------");
 cscanf ("%f",&km);
 cprintf("\nDistance between(3) and (4), mm----------");
 cscanf ("%f",&dist);
 cprintf("\nmv reading for a temp diff of 40 C-------");
 cscanf ("%f",&mvc);
 cprintf("\nTC1 , mv--------------------------------");
 cscanf ("%f",&tc1);
 cprintf("\nTC2 ,mv --------------------------------");
 cscanf ("%f",&tc2);
 cprintf("\nTC3 ,mv--------------------------------");
 cscanf ("%f",&tc3);
 cprintf("\nTC4 ,mv--------------------------------");
```

```
        cscanf ("%f",&tc4);


/* calculations*/

 /* bdh=20.0;
  diab=30.0;
  diac=10.0;
  dp=3.0;
  km=330.0;
  dist=25.0;
  mvc=40.0;
  tc1=5.0;
  tc2=4.0;
  tc3=3.6;
  tc4=3.2; */

  pi=3.141592;
  t1=tc1*mvc;
  t2=tc2*mvc;
  t3=tc3*mvc;
  t4=tc4*mvc;

  keff=(km*bdh/dist)*(pow(diac,2)/pow(diab,2))*((t3-t4)/(t1-t2));
  a=pi*pow((diab/1000.),2)/4.;
  q=keff*a*(t1-t2)/(bdh/1000.);

  printf("\n          RESULTS");
  printf("\n TC   Temp C");
  printf("\n TC1 temp , C %8.5f\n",t1);
  printf("\n TC2 temp , C %8.5f\n",t2);
  printf("\n TC3 temp , C %8.5f\n",t3);
  printf("\n TC4 temp , C %8.5f\n",t4);
  printf("\n Effective therm cond. of packed bed, kcal/hm %8.5f\n",keff);
  printf("\n Heat transfer rate through rod, kcal/h m2    %8.5f\n",q);

  fclose(fp);
  return 0;
  }
```

## 6.Heat transfer by free convection

```c
# include <stdio.h>
# include <math.h>
# include <conio.h>

 FILE *fp;


    float wid,lng,thk,cpa,cpm,kair,amb,ti,mvc,roa,rom,mua,vert;
    float tc1,tc2,tc3,tc4,cen,tcav,tavg,vol,area,beta,delt,nu;
    float gr,pr,grpr,nus,ha,x,y;

    int main(void)
    {
    window(10,10,100,30);
    textcolor(GREEN);
    textbackground(BLACK);

    fp=fopen("Exp6.txt","w");


    printf("6.0 HEAT TRANSFER BY FREE CONVECTION----- \n");

    printf("DATA INPUT\n");
    cprintf("Plate width, mm, mm------------------");
    cscanf ("%f",&wid);
    cprintf("\nPlate length, mm--------------------");
    cscanf ("%f",&lng);
    cprintf("\nPlate thickness, mm----------------");
    cscanf ("%f",&thk);
    cprintf("\nSp.heat. of air, kcal/kg C----------");
    cscanf ("%f",&cpa);
    cprintf("\nSp. heat of material, kcal/kg C-----");
    cscanf ("%f",&cpm);
    cprintf("\nTherm. cond. of air, kcal/hm C------");
    cscanf ("%f",&kair);
    cprintf("\nAmbient air temp. C----------------");
    cscanf ("%f",&amb);
    cprintf("\nInitial temp of plate, C------------");
    cscanf ("%f",&ti);
    cprintf("\nDeg C per mv ----------------------");
    cscanf ("%f",&mvc);
    cprintf("\nDensity of air, kg/m3---------------");
    cscanf ("%f",&roa);
    cprintf("\nDensity of meterial, kg/m3----------");
```

```
cscanf ("%f",&rom);
cprintf("\nViscosity of air, kg/m h -----------");
cscanf ("%f",&mua);
cprintf("\nTC reading corner 1 mv--------------");
cscanf ("%f",&tc1);
cprintf("\nTC reading corner 2 mv--------------");
cscanf ("%f",&tc2);
cprintf("\nTC reading corner 3 mv--------------");
cscanf ("%f",&tc3);
cprintf("\nTC reading corner 4 mv--------------");
cscanf ("%f",&tc4);
cprintf("\nTC reading centre   mv--------------");
cscanf ("%f",&cen);
cprintf("\nPlate orientation 1[vert],0[hoz]]---");
cscanf ("%f",&vert);

/* wid=20.0;
 lng=50.0;
 thk=3.0;
 cpa=0.24;
 cpm=0.20;
 kair=0.022;
 amb=26.;
 ti=40.0;
 mvc=40.0;
 roa=1.12;
 rom=8400.;
 mua=0.0694;

 tc1=1.25;
 tc2=1.24;
 tc3=1.249;
 tc4=2.51;
 cen=2.5;   */
/* calculations*/
 tcav=(tc1+tc2+tc3+tc4+cen)/5.;
 tavg=mvc * tcav;
 vol=wid*lng*thk/10e6;
 area=wid *lng/10e4;
 beta=1./(amb+273.);
 delt=(tavg-amb);
 nu=mua /(roa *3600.0);
 gr=9.8 * beta *(tavg-amb) *pow((lng/1000.0),3)/pow(nu,2);
 pr=cpa*mua/kair;
 grpr=gr*pr;
```

```c
if
(vert == 1)
{
nus = 0.521 * pow(grpr,0.5);
ha=nus * kair/(lng/1000.0);
}


if(vert == 0)
{

if(grpr>2.0e7 || grpr<=3.0e10);
{
nus =0.14*pow(grpr,0.33);
ha=nus*kair/(lng/1000.0);
}

if(grpr<2.0e7 || grpr>=10.0e5);
{
nus =0.54*pow(grpr,0.25);
ha=nus*kair/(lng/1000.0);
}

{
if(grpr<3.0e10 || grpr>3.0e5);
nus =0.27* pow(grpr,0.25);
ha=nus*kair/(lng/1000.0);
}
}

x=(ha*area)/(rom*cpm*vol);
y=(amb-tavg)/(amb-ti);


printf("\n                RESULTS");
printf("\n Plate are, cm2-------------------------%8.5f\n",area);
printf("\n Plate volume, cm3--------------------- %8.5f\n",vol);
printf("\n Beta coeff vol expans 1/K ------------ %8.5f\n",beta);
printf("\n Temp. of plate at any time,(Tavg),C--- %8.5f\n",tavg);
printf("\n Ngr ---- %8.5f\n",gr);
printf("\n Nnu -----%8.5f\n",nus);
printf("\n Npr ---- %8.5f\n",pr);
printf("\n Ngrpr--- %8.5f\n",grpr);
printf("\n ha calculated Eq 6.3, kcal/hm2 C---- %8.5f\n",ha);
printf("\n       VALUES FOR GRAPH");
```

```
    printf("\n (ha A / ro cp V)--------%8.5f\n",x);
    printf("\n (Too - T)/(Too - To)--- %8.5f\n",y);



/*  fclose(fp); */
 return 0;
 }
```

# 7.Double pipe heat exchanger

```c
# include <stdio.h>
# include <math.h>
# include <conio.h>

FILE *fp;

float id,od,lng,km,wat,tim,tc1,tc2,cpw,row,muw,kw,pres,tc3;
float vap,kc,roc,muc,mvc,pi;
float ca,flow,wvel,re,pr,hit,ho,dlm,c4,uo1,uo,vm8,tin,tout,ts;
float a1,a2,hi;
double delt,kc1,roc1,muc1,od1;

int main(void)
{
window(10,10,100,30);
textcolor(GREEN);
textbackground(BLACK);

fp=fopen("Exp7.txt","w");


printf("6.0 DOUBLE PIPE HEAT EXCHANGER----- \n");

printf("DATA INPUT\n");
cprintf("Inside diameter of the tube, mm-------");
cscanf ("%f",&id);
cprintf("\nOut side diameter of the tube, mm---");
cscanf ("%f",&od);
cprintf("\nLength of the tube, m--------------");
cscanf ("%f",&lng);
cprintf("\nTube material therm cond.kcak/hmC---");
cscanf ("%f",&km);
cprintf("\nWater collected, kg----------------");
cscanf ("%f",&wat);
cprintf("\nTime of collection , min------------");
cscanf ("%f",&tim);
cprintf("\nWater inlet TC mV------------------");
cscanf ("%f",&tc1);
cprintf("\nWater out let TC , mV--------------");
cscanf ("%f",&tc2);
cprintf("\nWater sp.heat, kcal/kg C------------");
cscanf ("%f",&cpw);
cprintf("\nWater density, kg/m3---------------");
cscanf ("%f",&row);
```

```
cprintf("\nViscosity of water, kg/mh-----------");
cscanf ("%f",&muw);
cprintf("\nTHerm cond water, kcal/hm C---------");
cscanf ("%f",&kw);
cprintf("\nSteam pressure, kN/m2--------------");
cscanf ("%f",&pres);
cprintf("\nSteam chest TC mV------------------");
cscanf ("%f",&tc3);
cprintf("\nLatent heat of vap, kcal/kg---------");
cscanf ("%f",&vap);
cprintf("\nCondensate therm cond, kcal/hm C----");
cscanf ("%f",&kc);
cprintf("\nCondensate density, kg/m3-----------");
cscanf ("%f",&roc);
cprintf("\nCondensate viscosity, kg/mh---------");
cscanf ("%f",&muc);
cprintf("\nDeg C for 1 mV value---------------");
cscanf ("%f",&mvc);

/*  id=14.0;
 od=22.0;
 lng=1.5;
 km=300.0;
 wat=9.14;
 tim=60.0;
 tc1=0.75;
 tc2=1.0;
 cpw=1.0;
 row=980.0;
 muw=1.44;
 kw=0.5055;
 pres=1.5;

 tc3=2.51;
 vap=538.0;
 kc=0.5849;
 roc=968.;
 muc=1.26;
 mvc=40.0; */
/* calculations*/


pi=3.141592;
tin=mvc*tc1;
tout=mvc*tc2;
ts=mvc*tc3;
```

```
ca=pi*(id/1000.)*(id/1000.0)/4.;
flow=wat*3600.0/tim;
wvel=flow/(row*ca*3600.0);
re=((id/1000.0)*wvel*row*3600.0)/muw;
pr=cpw*muw/kw;
hi=0.023*kw*1000.*pow(re,0.8)*pow(pr,0.4)/id;
delt=(ts-(tin+tout)/2.);
kc1=kc/3600.;
muc1=muc/3600.0;
od1=od/1000.0;
a1=pow(kc1,3)*vap*pow(roc,2)*9.81;
a2=(od1*muc1*delt);
ho= 0.725*pow((a1/a2),0.25)*3600.;
dlm=(od-id)/log(od/id);
c4=(((od-id)/1000.0)/km)*(od/dlm);

uo1=(1./ho) +c4 +od/(id*hi);
uo=1.0/uo1;
vm8=pow(wvel,0.8);



printf("\n                RESULTS");
printf("\n Water flow ratee,kg/h------------------%8.5f\n",flow);
printf("\n Water velocity in tube , m/s---------- %8.5f\n",wvel);
printf("\n Water inlet temp, C ------------------ %8.5f\n",tin);
printf("\n Water out let temp C------------------ %8.5f\n",tout);
printf("\n Steam condensing temp, C-------------- %8.5f\n",ts);
printf("\n ho condensate h t c, kcal/hm2 C------- %8.5f\n",ho);

printf("\n Nre ---- %8.5f\n",re);
printf("\n Npr -----%8.5f\n",pr);
printf("\n hi tube side h t c , kcal/hm2 C--- %8.5f\n",hi);
printf("\n For Wilson Plot\n");
printf("\n Vm -0.8 --- %8.5f\n",vm8);
printf("\n 1./Uo ------%8.5f\n",uo);
printf("\n 1/Uo = C3 * C4 * C5* Vm (0.8)\n");
printf("\n C4--- %8.8f\n",c4);



fclose(fp);
return 0;
}
```

Blank Page

## 8.Finned tube heat exchanger

```c
# include <stdio.h>
# include <math.h>
# include <conio.h>

FILE *fp;

float sflow,ts1,ts2,sps,mus,ks,ros,fo,leng,opid,ipod,ipid,km;
float pi,nfin,finh,finth,kfin,q,wat,lmtd,aa,gair,svel,res,prs;
float htcs,finef,afao,ai,pres,ts,roc,muc,vap,htc;
float finobs,ipca,opca,gan,wetp,de,ga,mutafaoai,muby,htct,p;
float omg,am,ab,hf,m,b,tanhmb,mb,xw,uo1,uo,dlm,ao1;
float a1,a2,a3,delt,hcon,af,ao,ai1,e1,e2;
float kc,m1,m2,m3;

int main(void)
{
window(10,10,100,30);
textcolor(GREEN);
textbackground(BLACK);

fp=fopen("Exp8.txt","w");


printf("8.0 FINNED TUBE HEAT EXCHANGER   \n");
printf(" (Double pipe longitudinal flow) \n");
printf("DATA INPUT\n");
cprintf("Air flow, kg/h------------------------");
cscanf ("%f",&sflow);
cprintf("\nIn let temp C----------------------");
cscanf ("%f",&ts1);
cprintf("\nOut let temp C---------------------");
cscanf ("%f",&ts2);
cprintf("\nAir sp.ht , kcal/kg C--------------");
cscanf ("%f",&sps);
cprintf("\nAir viscosity, kg/mh----------------");
cscanf ("%f",&mus);
cprintf("\nAir thermal cond. kcal/hm C---------");
cscanf ("%f",&ks);
cprintf("\nAir density, kg/m3------------------");
cscanf ("%f",&ros);
cprintf("\nFouling facto shel side, hmC/kcal---");
cscanf ("%f",&fo);
cprintf("\nPipe length, mm---------------------");
cscanf ("%f",&leng);
```

```
cprintf("\nOuter pipe ID, mm------------------");
cscanf ("%f",&opid);
cprintf("\nInner pipe OD, mm------------------");
cscanf ("%f",&ipod);
cprintf("\nInner pipe ID ,mm------------------");
cscanf ("%f",&ipid);
cprintf("\nPipe therm cond, kcal/h m C---------");
cscanf ("%f",&km);
cprintf("\nNo of fins-------------------------");
cscanf ("%f",&nfin);
cprintf("\nFin height, mm---------------------");
cscanf ("%f",&finh);
cprintf("\nFin thickness, mm------------------");
cscanf ("%f",&finth);
cprintf("\nFin material therm cond, kcal/hm C--");
cscanf ("%f",&kfin);
cprintf("\nSteam pressure, kN/m2--------------");
cscanf ("%f",&pres);
cprintf("\nSteam temp. C----------------------");
cscanf ("%f",&ts);
cprintf("\nCondensate therm cond, kcal/hm C----");
cscanf ("%f",&kc);
cprintf("\nCondensate density, kg/m3-----------");
cscanf ("%f",&roc);
cprintf("\nCondensate viscosity, kg/mh---------");
cscanf ("%f",&muc);
cprintf("\nLatent heat of vap, kcal/kg---------");
cscanf ("%f",&vap);


/* sflow=500.;
ts1=30.0;
ts2=40.;
sps=0.24;
mus=2.16;
ks=0.0265;
ros=0.961;
fo=0.0005;
leng=1.20;
opid=80.0;
ipod=40.;
ipid=34.;
km=30.;

nfin=15.;
finh=15.;
```

```
    finth=1.;
    kfin=225.;
    pres=1.5;
    ts=109.;
    kc=0.5847;
    roc=968.;
    muc=1.09;
    vap=538.0;  */

/* calculations*/

    pi=3.141592;
    q=sflow*sps*(ts2-ts1);
    wat=q/vap;
    lmtd=((ts-ts1)-(ts-ts2))/log((ts-ts1)/(ts-ts2));
    finobs=nfin*(finth/1000.0)*(finh/1000.0);
    ipca=pi*pow((ipod/1000.0),2)/4.;
    opca=pi*pow((opid/1000.0),2)/4.;
    aa=opca-ipca-finobs;
    gan=sflow/aa;
    svel=gan/(ros*3600.0);
    wetp=pi*(opid/1000.0)+nfin*(finth/1000.0)*2+ nfin*(finth/1000.0)
        -nfin*(finth/1000.0);
    de=4.*aa/wetp;
    res=de*gan/mus;
    prs=sps*mus/ks;


/*    heat transfer coeffshell side */

    htcs=(0.027*ks/de)*pow(res,0.8)*pow(prs,0.33);

    m1=htcs*2.0*nfin;
    m2=kfin*finth*nfin/1000.0;
    m3=m1/m2;
    m=pow(m3,0.5);
    mb=m*finh/1000.0;
    float  tanhmb=tanh(mb);
    finef=tanhmb/mb;

/*  Steam condensing coefficient */

    delt=(ts-(ts1+ts2)/2.);
    a1= pow((kc/3600.0),3)*vap*pow(roc,2)*9.81;
    a2=(ipid/1000.)*(muc/3600.0)*delt;
```

```
a3=a1/a2;
hcon=0.725*pow(a3,0.25)*3600.;


xw=(ipod-ipid)/(2.*1000.0);
dlm=((ipod-ipid)/log(ipod/ipid))/1000.;
af=nfin*finh*leng*2./1000.;

ao=pi*ipod*leng/1000.;

ai=ao*ipid/ipod;
afao=af+ao;
ao1=finef*af+ao;
uo1=ao1/(hcon*ai)+(xw/km)*(ao/dlm)+1./htcs+fo;
uo=1./uo1;




printf("\n                 RESULTS");
printf("\n Heat added , kcal/h--------------------%8.5f\n",q);
printf("\n Quantity of condensate, kg/h---------- %8.5f\n",wat);
printf("\n Lmtd C------------- ------------------ %8.5f\n",lmtd);
printf("\n Air flow area, m2--------------------- %8.5f\n",aa);
printf("\n Air mass velocity, kg/hm2------------- %8.5f\n",gan);
printf("\n Air velocity, m/s--------------------- %8.5f\n",svel);
printf("\n                 Shell side");
printf("\n Nre ---- %8.5f\n",res);
printf("\n Npr -----%8.5f\n",prs);
printf("\n Heat transfer coeff, kcal/hm2C-------- %8.5f\n",htcs);
printf("\n Fin efficiency fraction ---------------%8.5f\n",finef);
printf("\n Af Fins area, m2----------- %8.5f\n",af);
printf("\n Ao Bare tube area, m2------%8.5f\n",ao);
printf("\n Af + Ao area, m2, ---------%8.5f\n",afao);
printf("\n Ai tube inside area, m2--- %8.8f\n",ai);
printf("\n Condensation h t c, kcal/hm2C-------- %8.5f\n",hcon);
printf("\n Over all h t c , kcal/hm2C----------- %8.5f\n",uo);

fclose(fp);
return 0;
}
```

## 9.. Shell and Tube Heat Exchanger

```c
# include <stdio.h>
# include <math.h>
# include <conio.h>

FILE *fp;

float wat,ts1,ts2,spt,mut,kt,rot,fo,nt,tid,leng,pitc,shid,km;
float pres,ts,kc,roc,muc,vap,q,cnd,lmtd,ret,prt,htc,tca,tvel;
float uo,uoe,hcon,pi;


int main(void)
{
window(10,10,100,30);
textcolor(GREEN);
textbackground(BLACK);

 fp=fopen("Exp9.txt","w");


printf("9.0 SHELL AND TUBE HEAT EXCHANGER  \n");
printf(" ---------------------------------\n");
printf("DATA INPUT\n");
cprintf("Water flow,kg/h--------------------");
cscanf ("%f",&wat);
cprintf("\nIn let temp C--------------------");
cscanf ("%f",&ts1);
cprintf("\nOut let temp C-------------------");
cscanf ("%f",&ts2);
cprintf("\nAWater sp.ht , kcal/kg C----------");
cscanf ("%f",&spt);
cprintf("\nWater viscosity, kg/mh------------");
cscanf ("%f",&mut);
cprintf("\nWater thermal cond. kcal/hm C-----");
cscanf ("%f",&kt);
cprintf("\nWater density, kg/m3--------------");
cscanf ("%f",&rot);
cprintf("\nFouling facto shel side, hmC/kcal-");
cscanf ("%f",&fo);
cprintf("\nNo of tubes----------------------");
cscanf ("%f",&nt);
cprintf("\nTube ID, mm----------------------");
cscanf ("%f",&tid);
cprintf("\nTube length, mm------------------");
```

```
cscanf ("%f",&leng);
cprintf("\nTubes pitch (triangular),mm-------");
cscanf ("%f",&pitc);
cprintf("\nShell ID, mm--------------------");
cscanf ("%f",&pitc);
cprintf("\nTubee therm cond, kcal/h m C------");
cscanf ("%f",&km);
cprintf("\nSteam pressure, kg/cm2g-----------");
cscanf ("%f",&pres);
cprintf("\nSteam temp. C--------------------");
cscanf ("%f",&ts);
cprintf("\nCondensate therm cond, kcal/hm C--");
cscanf ("%f",&kc);
cprintf("\nCondensate density, kg/m3---------");
cscanf ("%f",&roc);
cprintf("\nCondensate viscosity, kg/mh-------");
cscanf ("%f",&muc);
cprintf("\nLatent heat of vap, kcal/kg-------");
cscanf ("%f",&vap);


/*  wat=3000.;
 ts1=30.0;
 ts2=40.;
 spt=1.0;
 mut=0.0772;
 kt=0.0238;
 rot=980.0;
 fo=0.0005;
 nt=12.;
 tid=10.0;
 leng=600.;
 pitc=25.;
 shid=150.;
 km=40.;
 pres=1.5;
 ts=109.;
 kc=0.5847;
 roc=968.;
 muc=1.09;
 vap=538.0;    */

/* calculations*/
  float cnd,delt,a1,a2,a3,uo1;
  pi=3.141592;
  q=wat*spt*(ts2-ts1);
```

```
cnd=q/vap;
lmtd=((ts-ts1)-(ts-ts2))/log((ts-ts1)/(ts-ts2));
/* tube side  */
tca=nt*pi*pow((tid/1000.0),2)/4.;
tvel=wat/(rot*tca*3600);
ret=(tid/1000.0)*tvel*3600.*rot/mut;
prt=spt*mut/kt;

htc=(0.027*kt*1000./tid)*pow(ret,0.8)*pow(prt,0.33);

/*  Steam condensing coefficient */

delt=(ts-(ts1+ts2)/2.);
a1= pow((kc/3600.0),3)*vap*pow(roc,2)*9.81;
a2=(tid/1000.)*(muc/3600.0)*delt;
a3=a1/a2;
hcon=0.725*pow(a3,0.25)*3600.;

/* overall htc  */
uo1=1./hcon + 1./htc + fo;
uo=1./uo1;

 /* Eq 9.4    */

float ao;
ao=nt*pi*tid*leng/10e6;
uoe=wat*spt*(ts2-ts1)/(ao*lmtd);


printf("\n               RESULTS");
printf("\n Heat added , kcal/h--------------------%8.5f\n",q);
printf("\n Quantity of condensate, kg/h---------- %8.5f\n",cnd);
printf("\n Lmtd C------------ ------------------ %8.5f\n",lmtd);
printf("\n Tube side");
printf("\n Nre ---- %8.5f\n",ret);
printf("\n Npr -----%8.5f\n",prt);
printf("\n Heat transfer coeff, kcal/hm2C-------- %8.5f\n",htc);
printf("\n Water flow area, m2--------------------%8.5f\n",tca);
printf("\n Water velocity, m/s------------------- %8.5f\n",tvel);
printf("\n Uo Overall h t c ,kcal/hm2C------------%8.5f\n",uo);
printf("\n Uo from Eq 9.4, kcal/hm2 C-------------%8.5f\n",uoe);
printf("\n Condensing h t c ,kcal/hm2 C---------- %8.8f\n",hcon);

fclose(fp);
return 0;
}
```

Blank Page

# 10.Heat transfer in agitated vessels

```c
# include <stdio.h>
# include <math.h>
# include <conio.h>

 FILE *fp;

 float cid,cod,hd,ncoil,leg1,leg2,vid,pdl,rpm,tin,tout,tb,row;
 float muw,kw,kcoil,cpw,muww,wat,wth,mdia,stlnth,ai,ao,ma,ca,tvel;
 float re,pr,htc,dtm,hic,hto,uo,pi;


 int main(void)
 {
 window(10,10,100,30);
 textcolor(GREEN);
 textbackground(BLACK);

  fp=fopen("Exp10.txt","w");


 printf("10.0 HEAT TRANSFER IN AGITATED VESSELS  \n");
 printf(" --------------------------------\n");
 printf("DATA INPUT\n");
 cprintf("Inside diameter of coil,mm----------");
 cscanf ("%f",&cid);
 cprintf("\nlOutside diameter of coil,mm------");
 cscanf ("%f",&cod);
 cprintf("\nDiameter of helix, mm-------------");
 cscanf ("%f",&hd);
 cprintf("\nNo of coils----------------------");
 cscanf ("%f",&ncoil);
 cprintf("\nVertical leg1 immersed in bath, mm");
 cscanf ("%f",&leg1);
 cprintf("\nVertical leg2 immersed in bath, mm");
 cscanf ("%f",&leg2);
 cprintf("\nInternal diameter of vessel,mm----");
 cscanf ("%f",&vid);
 cprintf("\nLength of paddle, mm--------------");
 cscanf ("%f",&pdl);
 cprintf("\nNo of revelutions permin, rpm-----");
 cscanf ("%f",&rpm);
 cprintf("\nWater inlet temp, C--------------");
 cscanf ("%f",&tin);
 cprintf("\nWater outlet temp C--------------");
```

```
cscanf ("%f",&tout);
cprintf("\nBath temperature, C---------------");
cscanf ("%f",&tb);
cprintf("\nDensity of water, kg/m3-----------");
cscanf ("%f",&row);
cprintf("\nViscosity of water, kg/mh---------");
cscanf ("%f",&muw);
cprintf("\nThermal cond. of water, kcal/hm C-");
cscanf ("%f",&kw);
cprintf("\nCoil wall meterial k, kcal/hm C---");
cscanf ("%f",&kcoil);
cprintf("\nSp. heat of water, kcal/kg C------");
cscanf ("%f",&cpw);
cprintf("\nVisc. of water at wall temp,kg/mh-");
cscanf ("%f",&muww);
cprintf("\nMass flow rate of water, kg/h------");
cscanf ("%f",&wat);

/* cid=22.0;
 cod=25.0;
 hd=800.0;
 ncoil=20.0;
 leg1=170.;
 leg1=670.;
 vid=1000.0;
 pdl=300.0;
 rpm=800.0;
 tin=30.0;
 tout=40.0;
 tb=100.0;
 row=980.0;
 muw=0.072;
 kw=0.0236;
 kcoil=30.0;
 cpw=1.0;
 muww=0.06;
 wat=1400.0; */

/* calculations*/

 pi=3.141592;
 wth=(cod-cid)/2.;
 mdia=(cod+cid)/2.;
 stlnth=leg1+leg2+pi*hd*ncoil;
 ai=pi*cid*stlnth/1e6;
 ao=pi*cod*stlnth/1e6;
```

```
ma=(ai+ao)/2.;
ca=pi*pow((cid/1000.0),2)/4.;
tvel=wat/(row*ca*3600.0);
re=cid*tvel*row*3.6/muw;
pr=cpw*muw/kw;

/*  Inside HTC equations  */
htc=(0.023*kw*1000./cid)*pow(re,0.8)*pow(pr,0.4);
hic=htc*(1.0+3.5*cid/mdia);

/*  Outside HTC equation    */
float cons1,cons2,cons3,cons4,u2o,u2;
cons1=((vid/1000.0)/kw)*pow((muw/muww),0.14);
cons2=pow(pr,0.3333);
cons4=pow((pdl/1000.0),2);
cons3=pow((cons4*rpm*60.*row/muw),0.62);
hto=0.87*cons2*cons3/cons1;
dtm=((tb-tin)-(tb-tout))/log((tb-tin)/(tb-tout));
uo=wat*cpw*(tout-tin)/(ao*dtm);
u2o=cod/(cid*htc)+(wth/(1000.*kcoil))*(cod/cid)+1/hto;
u2=1./u2o;

printf("\n                 RESULTS");
printf("\n Coil wall thickness, mm----------------%8.5f\n",wth);
printf("\n Mean dia. of pipe wall, mm/h---------- %8.5f\n",mdia);
printf("\n Length of straight coil, mm----------- %8.5f\n",stlnth);
printf("\n Coil inside surface area, m2---------- %8.5f\n",ai);
printf("\n Coil outside surface area, m2----------%8.5f\n",ao);
printf("\n Mean of inside and outside area, m2--- %8.5f\n",ma);
printf("\n Pipe cross sectional area, m2----------%8.5f\n",ca);
printf("\n Water velocity in coil, m/s------------ %8.5f\n",tvel);
printf("\n Reynolds number------------------------%8.5f\n",re);
printf("\n Prandtl number-------------------------%8.5f\n",pr);
printf("\n Inside film h t c ,kcal/hm2 C--------- %8.8f\n",htc);
printf("\n Lmtd, C------------------------------ %8.5f\n",dtm);
printf("\n hi value eq 10.4, kcal/hm2 C-----------%8.5f\n",hic);
printf("\n ho value eq 10.5, kcal/m2 C------------%8.5f\n",hto);
printf("\n Uo value eq 10.7, kcal/hm2 C---------- %8.8f\n",u2);
printf("\n Uo value eq 10.6, kcal/hm2 C---------- %8.8f\n",uo);
fclose(fp);
return 0;
}
```

Blank Page

# 11. Heat transfer to boiling liquids

```c
# include <stdio.h>
# include <math.h>
# include <conio.h>

 FILE *fp;

 float dia,lnth,hsa,tb,vap,mu,ro,sig,cp,stmro,kw,vl,il,tbase;
 float spro,tcor,ca,ohm,w,delt,htc,ohm1,tw,wa,pi,flux,qbya;



 int main(void)
 {
 window(10,10,100,30);
 textcolor(GREEN);
 textbackground(BLACK);

  fp=fopen("Exp11.txt","w");



 printf("11.0 HEAT TRANSFER TO BOILING LIQUIDS  \n");
 printf(" ---------------------------------\n");
 printf("DATA INPUT\n");
 cprintf("diameter of wire, mm---------------");
 cscanf ("%f",&dia);
 cprintf("\nLength of wire, mm---------------");
 cscanf ("%f",&lnth);
 cprintf("\nWater bath temp C----------------");
 cscanf ("%f",&tb);
 cprintf("\nLatent heat of vapo.kcal/kg-------");
 cscanf ("%f",&vap);
 cprintf("\nWater viscosity, kg/mh------------");
 cscanf ("%f",&mu);
 cprintf("\nWater density, kg/m3--------------");
 cscanf ("%f",&ro);
 cprintf("\nWater surface tension , kg/m------");
 cscanf ("%f",&sig);
 cprintf("\nWater sp. heat ,kcal/kg C---------");
 cscanf ("%f",&cp);
 cprintf("\nSteam density, kg/m3--------------");
 cscanf ("%f",&stmro);
 cprintf("\nWater therm cond, kcal/hm C-------");
 cscanf ("%f",&kw);
 cprintf("\nPlatinum wire base temp, C--------");
```

```
cscanf ("%f",&tbase);
cprintf("\nPl. resitivity, ohm/m2 x 1e-8-----");
cscanf ("%f",&spro);
cprintf("\nTemp. coeff. of resitance, ohm/C--");
cscanf ("%f",&tcor);
printf("\nTABULATE THE VALUES FOR\n ");

cprintf("\nVolts V-------------");
cscanf ("%f",&vl);
cprintf("\n Current, I----------");
cscanf ("%f",&il);

/*  dia=0.25;
 lnth=150.0;
 tb=90.0;
 vap=583.0;
 mu=1.021;
 ro=968.0;
 sig=0.058;
 cp=1.0;
 stmro=5.975;
 kw=0.026;
 vl=1.0;
 il=3.0;
 tbase=100.0;
 spro=11.;
 tcor=0.0039;     */

/* calculations*/

 pi=3.141592;
 ca=pi*dia*dia*1.e-6/4.;
 ohm=spro*1e-8*lnth/(ca*1000.);
 hsa=pi*dia*lnth*1.e-6;
 wa=vl*il;
 ohm1=vl/il;
 tw=tbase+tbase*(1.0+(ohm1-ohm)/tcor);
 delt=tw-tb;
 htc=wa*0.86/(hsa*delt);
 flux=wa*0.86/hsa;
 float cons2,cons1,cons3,cons4,cons5,qa13,qbya;
 cons2=cp*mu/kw;
 cons1=cp*delt/(vap*cons2);
 cons3= pow(sig/((ro-stmro)*9.8),0.5);
 cons4=vap*mu;
 cons5=pow((cons3/cons4),0.333);
```

```
qa13=cons1/(0.013*cons5);
qbya = pow(qa13,3);




printf("\n                RESULTS");
printf("\n Wire heat transfer ara, m2-------------%8.5f\n",hsa);
printf("\n Wire resitance, ohm-------------------- %8.5f\n",ohm1);
printf("\n Wire surface temperature, C----------- %8.5f\n",tw);
printf("\n Wire temp - bath temp, C-------------- %8.5f\n",delt);
printf("\n Heat transfer rate, W-------------------%8.5f\n",wa);
printf("\n Heat flux , kcal/m2-------------------- %8.5f\n",flux);
printf("\n H T C , kcal/hm2 C---------------------%8.5f\n",htc);
printf("\n H T C eq 11.1, kcal/m2 C-------------- %8.5f\n",qbya);

 fclose(fp);
return 0;
}
```

Blank Page

# 12.Heat transfer to gas fluidized beds

```c
# include <stdio.h>
# include <math.h>
# include <conio.h>

 FILE *fp;

 float air,htrd,htrh,dp,cd,bd,eps,cps,ros,kg,cpg,rog,mug,dmv;
 float ohm,amp,volt,wa,kcal,mvc,delt,cum,bca,gvel,mfg,htra,qflux;
 float hexp,rep,pr,htcf,hbed,pi,cons1,cons2,cons3,cons4;


 int main(void)
 {
 window(10,10,100,30);
 textcolor(GREEN);
 textbackground(BLACK);

  fp=fopen("Exp12.txt","w");


 printf("12. HEAT TRANSFER IN GAS FLUIDIZED BEDS  \n");
 printf(" --------------------------------\n");
 printf("DATA INPUT\n");
 cprintf("Air flow kg/h----------------------");
 cscanf ("%f",&air);
 cprintf("\nHeater diameter,mm----------------");
 cscanf ("%f",&htrd);
 cprintf("\nHeater height, mm-----------------");
 cscanf ("%f",&htrh);
 cprintf("\nParticle diameter, mm-------------");
 cscanf ("%f",&dp);
 cprintf("\nInside diameter of column, mm-----");
 cscanf ("%f",&cd);
 cprintf("\nHeight of bed, mm-----------------");
 cscanf ("%f",&bd);
 cprintf("\nBed voidage-----------------------");
 cscanf ("%f",&eps);
 cprintf("\nSolid particle sp.heat ,kcal/kg C-");
 cscanf ("%f",&cps);
 cprintf("\nDensity of solid , kg/m3----------");
 cscanf ("%f",&ros);
 cprintf("\nTherm cond. of gas, kcal/hmC------");
 cscanf ("%f",&kg);
```

```
cprintf("\nSp. heat of gas, kcal/kg C--------");
cscanf ("%f",&cpg);
cprintf("\nDensity of gas, kg/m3-------------");
cscanf ("%f",&rog);
cprintf("\nViscosity of gas, kg/m h----------");
cscanf ("%f",&mug);
cprintf("\nTC millivolts diffe ence, mV------");
cscanf ("%f",&dmv);
cprintf("\nHeater resitance, ohm-------------");
cscanf ("%f",&ohm);
cprintf("\nCurrent ,Amps--------------------");
cscanf ("%f",&amp);
cprintf("\nVoltage, Volts, ohm--------------");
cscanf ("%f",&volt);
cprintf("\nDeg C risec for 1 mV TC reading---");
cscanf ("%f",&mvc);



/* air=30.0;
htrd=13.0;
htrh=25.0;
dp=1.0;
cd=100.0;
bd=100.0;
eps=0.4;
cps=0.16;
ros=2600.0;
kg=0.0269;
cpg=0.24;
rog=0.969;
mug=0.0772;
dmv=0.115;
ohm=140.0;
amp=1.0;
volt=2.0;
mvc=40.0; */

/* calculations*/

pi=3.141592;
delt=dmv*mvc;
bca=pi*pow((bd/1000.0),2)/4.;
cum=air/rog;
gvel=cum/(bca*3600.0);
mfg=air/bca;
```

```c
wa=amp*volt;
kcal=0.86*wa;
htra=pi*htrd*htrh;
qflux=kcal*1e6/htra;
hexp=qflux/delt;
rep=(dp/1000.0)*mfg/mug;
pr=cpg*mug/kg;

if (rep < 350.0)
{
htcf=1.06*pow(rep,-0.41)*cpg*mfg/pow(pr,0.666);
}

if(rep >= 350. || rep <4000.0)
{
cons1=pow((cd/bd),0.65);
cons2=pow((cd/dp),0.17);
cons3=pow((((1.0-eps)*ros*cps)/(eps*rog*cpg)),0.25);
cons4=pow(((bd/1000.0)*mfg/mug),0.8);
htcf=0.55*cons1*cons2*cons3*cons4*kg/(cd/1000.0);
}




printf("\n               RESULTS");
printf("\n Electrical input, W--------------------%8.5f\n",wa);
printf("\n Electrical input, kcal/h-------------- %8.5f\n",kcal);
printf("\n Wall - Bed temp diff, C--------------- %8.5f\n",delt);
printf("\n Volumetric flow rate of air, m3/h----- %8.5f\n",cum);
printf("\n Bed cross sectional area , m2----------%8.5f\n",bca);
printf("\n Superficial gas velocity , m/s-------- %8.5f\n",gvel);
printf("\n Mass velocity of gas, kg/m2 h----------%8.5f\n",mfg);
printf("\n Heater surface area, mm2-------------- %8.5f\n",htra);

printf("\n Heat flux, kcal/hm2------------------- %8.5f\n",hexp);
printf("\n Rep particle reynolds number-----------%8.5f\n",rep);
printf("\n Prandtl number----------------------- %8.5f\n",pr);
printf("\n Heat tran. coef. eq 12.2 kcal/m2 h C---%8.5f\n",htcf);


  fclose(fp);
return 0;
}
```

Blank Page

## 13.Long tube vertical evaporator

```
# include <stdio.h>
# include <math.h>
# include <conio.h>

FILE *fp;

float pr,tod,tubl,xf,xp,tc1,tc2,tf,tp,cpf,cpw,vap,tc,feed,cw,pi;
float wa,cond,a,q,u;




int main(void)
{
window(10,10,100,30);
textcolor(GREEN);
textbackground(BLACK);

 fp=fopen("Exp13.txt","w");



printf("12. LONG TUBE VERTICAL EVAPOTATOR  \n");
printf(" ---------------------------------\n");
printf("DATA INPUT\n");
cprintf("Evaporator pressure, kg/cm2g---------------");
cscanf ("%f",&pr);
cprintf("\nMean dia. of the tube, mm----------------");
cscanf ("%f",&tod);
cprintf("\nEffective heat transfer length, m--------");
cscanf ("%f",&tubl);
cprintf("\nConcentration of solute in feed, mass%---");
cscanf ("%f",&xf);
cprintf("\nConcentration of solute in product, mass%");
cscanf ("%f",&xp);
cprintf("\nCooling water inlet temp, C--------------");
cscanf ("%f",&tc1);
cprintf("\nCooling water outlet temp , C------------");
cscanf ("%f",&tc2);
cprintf("\nTemperature of feed, C-------------------");
cscanf ("%f",&tf);
cprintf("\nTemperature of product, C---------------");
cscanf ("%f",&tp);
cprintf("\nSp.heat of feed , kcal/kg C--------------");
cscanf ("%f",&cpf);
```

```
cprintf("\nSp. heat of cooling water, kcal/kg C-----");
cscanf ("%f",&cpw);
cprintf("\nLatent heat of evaporation, kcal/kg------");
cscanf ("%f",&vap);
cprintf("\nTemperature of condensed vapor, C--------");
cscanf ("%f",&tc);
cprintf("\nMass flow rate of feed ,kg/h-------------");
cscanf ("%f",&feed);
cprintf("\nMass flow rate of cooling water, kg/h----");
cscanf ("%f",&cw);




/* calculations*/

  pi=3.141592;
  wa=feed*xf/xp;
  cond=cw*cpw*(tc2-tc1)/vap;
  q=(feed-wa)*(tp-tc)+feed*cpf*(tp-tf)+cw*cpw*(tc2-tc1);
  a=pi*tod*tubl/1000.;
  u=q/(a*(tc-tp));




  printf("\n                RESULTS");
  printf("\n Mass flow rate of product liquor, kg/h-%8.5f\n",wa);
  printf("\n Condensate rate, kg/h----------------- %8.5f\n",cond);
  printf("\n Heat transfer area, m2---------------- %8.5f\n",a);
  printf("\n Heat transfer to liquid, kcal/h------- %8.5f\n",q);
  printf("\n Overall H T C , kcal/hm2 C------------%8.5f\n",u);


  fclose(fp);
  return 0;
}
```

## 14. Radition constat

```c
# include <stdio.h>
# include <math.h>
# include <conio.h>

FILE *fp;

float pi,cid,cht,cps,ros,cpg,kg,rog,mug,amb,eps,mvc,tmp,mv,ts;
float a,v,mass,gr,pr,q,hr;




int main(void)
{
window(10,10,100,30);
textcolor(GREEN);
textbackground(BLACK);

 fp=fopen("Exp14.txt","w");


printf("14. THERMAL RADIATION CONSTANTS \n");
printf(" --------------------------------\n");
printf("DATA INPUT\n");
cprintf("Cylinder diameter, mm---------------------");
cscanf ("%f",&cid);
cprintf("\nCylinder height, mm---------------------");
cscanf ("%f",&cht);
cprintf("\nSp.heat of solid ,kcal/kg C--------------");
cscanf ("%f",&cps);
cprintf("\nDensity of solid ,kg/m3------------------");
cscanf ("%f",&ros);
cprintf("\nSp.heat of fluid,kcal/kg C---------------");
cscanf ("%f",&cpg);
cprintf("\nTherm cond. of fluid , kcal/hm C---------");
cscanf ("%f",&kg);
cprintf("\nDensity of fluid, kg/m3------------------");
cscanf ("%f",&rog);
cprintf("\nViscosity of fluid , kg/mh--------------");
cscanf ("%f",&mug);
cprintf("\nAmbient temp, C--------------------------");
cscanf ("%f",&amb);
cprintf("\nBody emissivity, kcal/hm2 K-1------------");
cscanf ("%f",&eps);
cprintf("\nDeg rise for 1mV Tc reading, Cl/kg C-----");
```

```
cscanf ("%f",&mvc);
cprintf("\nThermo couple reading, mV---------------");
cscanf ("%f",&tmp);




/*  cid=80.;
 cht=120.;
 cps=0.121;
 ros=8760.0;
 cpg=0.24;
 kg=0.0269;
 rog=0.969;
 mug=0.0772;
 amb=30.0;
 eps=0.8;
 mvc=40.;
 tmp=2.5;
 mv=5.;  */



/* calculations*/
 float muro,grpr,beta,b,n;
 pi=3.141592;
 a=pi*cid*cht/1.e6+2.*pi*pow((cid/1000.0),2)/4.;
 v=pi*pow((cid/1000.0),2)*(cht/1000.0)/4.;
 mass=v*ros;
 beta=1.0/(273.+amb);
 pr=cpg*mug/kg;

 muro=(mug/3600.0)/rog;
 ts=tmp*mvc;
 gr=9.8*beta*(ts-amb)*pow((cht/1000.0),3)/pow(muro,2);
 grpr=gr*pr;

 if(grpr<1.e4)
 {
 b=1.36;
 n=0.125;
 }
 if(grpr>1.e4 || grpr<1.e9)
 {
 b=0.59;
 n=0.25;
 }
```

```
if(grpr>1.e9)
{
b=0.13;
n=0.333;
}

hr=b*(kg*1000./cht)*pow(grpr,n);
q=hr*(ts-amb)*a;




printf("\n                 RESULTS");
printf("\n Solid surface area, m2---------------- %8.5f\n",a);
printf("\n Solid volume, m3---------------------- %8.5f\n",v);
printf("\n Mass of solid, kg----------------------%8.5f\n",mass);
printf("\n Solid temperature, C-------------------%8.5f\n",ts);
printf("\n Coeff of volume expasion of fluid, 1/K-%8.5f\n",beta);

printf("\n Grashoff number----------------------- %8.5f\n",gr);
printf("\n Prandtl number------------------------ %8.5f\n",pr);
printf("\n Radiation H T C, kcal/hm2 C------------%8.5f\n",hr);
printf("\n Heat transfer rate, kcal/--------------%8.5f\n",q);
printf("\n DRAW A GRAPH OF TEMPERATURE WITH TIME AND ");
printf("\n  GET THE VALUE OF dT/dt to solve Eq 14.6 ");

fclose(fp);
return 0;
}



;
```

Blank Page

# References

1. Coulson, J. M., and Richardson, J. F., "Chemical Engineering, Vol 1", Pergaman Press, Oxford 1976.

2. Kern, D. Q., "Process Heat Transfer ",
   Mc Graw Hill, New York, 1950.

3. Mc Adams, W. H., " Heat Transmission ", 3$^{rd}$ ed.
   Mc Graw Hill, New York, 1954.

4. Mc Cabe, W. L., and Smith, J. C. , "Unit operations in Chemical Engineering" , 3$^{rd}$ ed.
   Mc Graw Hill & Kogakusha,1976.

5. Platt, F and Kirchner, K. , "Praktikum der Technischen Chemie",
   Walter de Gruyter, Berlin, 1975.

6. Perry, J. H. , " Chemical Engineers Hand Book " 5$^{th}$ ed.
   Mc Graw Hill, New York,   latest Ed.

7. Sankaran, K. , " Experiments in Heat Transfer and Mass Transfer "
   Engineering college, Trichur-9

8. Papers Presented at " The conference on Chemical Engineering Education ',
   Indian Institute of Technology, Madras, Jan 1977.

9. N S Srinivas, Heat Transfer Operations, A laboratory manual
   Published by CDC , Indian Institute of Technology, Madras, Chennai
   Feb (1980)

Blank Page

# Appendix – factors for conversion to SI units

| Mass | |
|---|---|
| I lb | 0.454 kg |
| 1 ton | 1016. kg |

| Length | |
|---|---|
| I inch | 25.4 mm |
| 1 ft | 0.305 m |
| 1 mile | 1.609 km |

| Time | |
|---|---|
| 1 min | 60 s |
| 1 h | 3.6 ks |
| 1 day | 86.4Ms |

| Area | |
|---|---|
| 1 sq in | 645.2 sq mm |
| 1 sq ft | 0.093 sq m |

| Volume | |
|---|---|
| 1 cu in | 16,387.1 cu mm |
| 1 cu ft | 0.0283 cu m |
| 1 UK gal | 4546 cu cm |
| 1 US gal | 3786 cu cm |

| Force | |
|---|---|
| 1 poundal | 0.138 N |
| 1 pond force | 4.45 N |
| 1 dyne | $10^{-5}$ N |

| Energy | |
|---|---|
| 1 ft lbf | 1.36 J |
| 1 cal | 4.187 J |
| 1 erg | $10^{-7}$ J |
| 1 Btu | 1.055 kJ |

| Power | |
|---|---|
| 1 hp | 745 W |
| 1 Btu/h | 0.293 W |

## Appendix – factors for conversion to SI units (contd)

| Pressure | |
|---|---|
| 1 lbf.sq in | 6.895 kN/sq m |
| 1 atm | 101.3 kN/sq m |
| 1 bar | 100 kN/sq m |
| 1 ft water | 2.99 kN/sq m |
| 1 in water | 249 N/sq m |
| 1 in Hg | 3.39 kN/sq m |
| 1 mm Hg | 133 N/sq m |

| viscosity | |
|---|---|
| 1 P | 0.1 Ns/sq m |
| 1 lb/ft h | 0.414 Ns/sq m |
| 1 stoke | $10^{-7}$ sq m/s |
| 1 ft2/h | 0.258 sq cm/s |

| Mass flow | |
|---|---|
| 1 lb/h | 0.126 g/s |
| 1 ton/h | 0.282 kg/s |
| 1 lb/hft$^2$ | 1.356 g/s sq m |

| Thermal | |
|---|---|
| 1 btu/h ft$^2$ | 3.155 W/m$^2$ |
| 1 btu/h ft$^2$ °F | 5.678 W/m$^2$K |
| 1 btu/lb | 2.326 kJ/kg |
| 1 btu/lb °F | 4.187 kJ/kg K |
| 1 btu/h ft °F | 1.731 W/mK |

| Energy | |
|---|---|
| 1 kWh | 3.6 MJ |
| 1 therm | 105.5 MJ |

| Calorific value | |
|---|---|
| 1 btu/ft$^3$ | 37.26 kJ/m$^3$ |
| 1 btu/lb | 2.326 kJ/kg |

| Density | |
|---|---|
| 1 lb/ft$^3$ | 16.02 kg/m$^3$ |

www.ingramcontent.com/pod-product-compliance
Lightning Source LLC
Chambersburg PA
CBHW081724170526
45167CB00009B/3697